BIOLOGICAL AND MEDICAL PHYSICS, BIOMEDICAL ENGINEERING

BIOLOGICAL AND MEDICAL PHYSICS, BIOMEDICAL ENGINEERING

The fields of biological and medical physics and biomedical engineering are broad, multidisciplinary and dynamic. They lie at the crossroads of frontier research in physics, biology, chemistry, and medicine. The Biological and Medical Physics, Biomedical Engineering Series is intended to be comprehensive, covering a broad range of topics important to the study of the physical, chemical and biological sciences. Its goal is to provide scientists and engineers with textbooks, monographs, and reference works to address the growing need for information.

Books in the series emphasize established and emergent areas of science including molecular, membrane, and mathematical biophysics; photosynthetic energy harvesting and conversion; information processing; physical principles of genetics; sensory communications; automata networks, neural networks, and cellular automata. Equally important will be coverage of applied aspects of biological and medical physics and biomedical engineering such as molecular electronic components and devices, biosensors, medicine, imaging, physical principles of renewable energy production, advanced prostheses, and environmental control and engineering.

Ari Ide-Ektessabi

Applications of Synchrotron Radiation

Micro Beams in
Cell Micro Biology and Medicine

With 135 Figures, 4 in color

 Springer

Professor Ari Ide-Ektessabi
Kyoto University
International Innovation Center
Bio System Electronics
Yoshida Honmachi Sakyoku
606-8501 Kyoto, Japan
E-mail: h51167@sakura.kudpc.kyoto-u.ac.jp

Library of Congress Control Number: 2007920612

ISSN 1618-7210

ISBN 978-3-540-46424-2 Springer Berlin Heidelberg New York

Springer is a part of Springer Science+Business Media

springer.com

Typesetting and production: LE-TEX Jelonek, Schmidt & Vöckler GbR, Leipzig
Cover: eStudio Calamar Steinen

SPIN 11892724   57/3100/YL - 5 4 3 2 1 0   Printed on acid-free paper

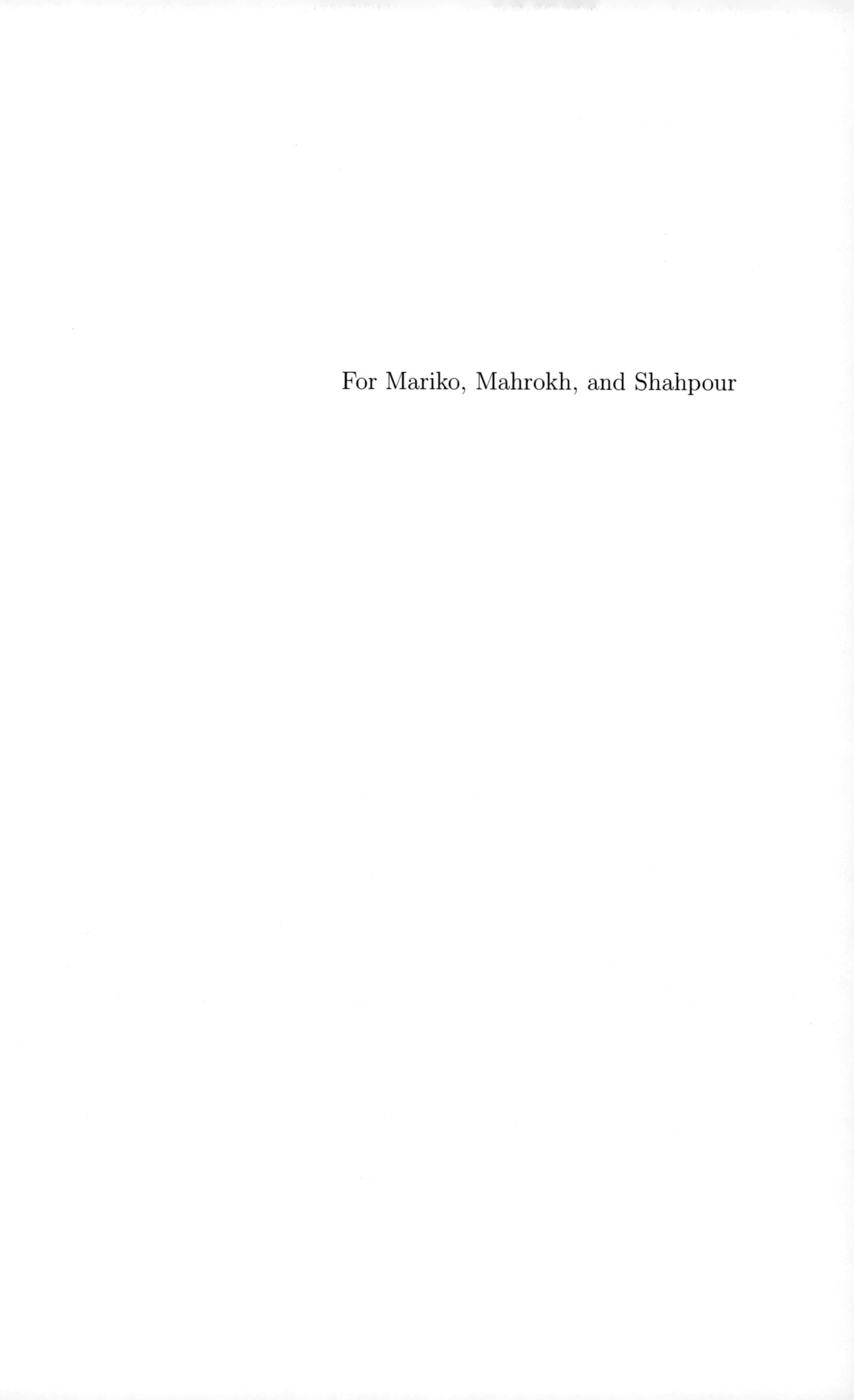

For Mariko, Mahrokh, and Shahpour

Preface

Physics and engineering governing the applications of synchrotron radiation is based on enormous achievements during more than one hundred years in the filed of X-ray physics and technology. The contents of this book, starting with the very general aspects of synchrotron radiation investigated, have been developed by numerous scientist and experimentalist in this field during the past 20 years. The readers are recommended to visit the websites of major synchrotron facilities in the world, and update their knowledge of this rapidly changing and progressing field. Since this book covers a wide range of topics related to the experimental aspects, from the physics to biology, in many occasions it does not cover the many important works by scientists in the field. A note of acknowledgement must begin with a sincere apology of shortcomings in referring to all the important works in the field.

The author is very much indebted to Professor Atsuo Iida of Photon factory, Tsukuba Japan, where most of the experiments and achievements reported in this book were initiated. I should express my thanks to all the researchers in High Energy Accelerator Research Organization, Institute of Material Structure Science (Photon Factory) and Japan Synchrotron Radiation Research Institute (SPring8), who made the experiments possible for me and my colleagues who performed the main experimental topics during 1995 to 2003 presented in this book. Contribution by Professor S. Hayakawa, in the preliminary experiments related to chemical-state imaging at a single cell level in SPring8 was very important in establishing the methodology shown in Chap. 6 of the book. I am very much indebted to Professor Soey Sie for his critical reading of the manuscript during his visit to our laboratory in 2005. I would like to express my thanks to Mariona Rabionet Roig for reading and assistance in preparation of the manuscript.

I am greatly indebted to Professor Sohei Yoshida and Dr. Ryoko Ishihara for their participation in experiments and discussions related to neurodegeneration, in this book. My graduate students, Kouji Takada, Shigeyoshi Fujisawa, Norio Kitamura, Shunsuke Shikine, Koyo Shirasawa, and Takuo Kawakami, did most of the experiments and investigations in this work. Without their contributions, this book could not have been written.

Contents

1

Introduction

The aim of this book is to demonstrate the applications of synchrotron radiation in certain aspects of cell microbiology, specifically non-destructive elemental analyses, chemical-state analyses and imaging (distribution) of the elements within a cell. The basics for understanding and applications of synchrotron radiation are the same as those of X-ray spectrometry, which has been well developed during the twentieth century and is widely applied to various fields of science and technology, including biology and medicine. The two main techniques that will be discussed in this book are the X-ray fluorescence spectroscopy (XRF) and the X-ray fine structure analysis (XAFS). What makes a synchrotron radiation X-ray source very useful for analytical works, especially for biological applications, are the very high brilliance and energy variability of the X-ray beam.

X-rays from a synchrotron radiation source enable non-destructive determination of the distribution and the chemical state (the electronic structure of the elements) of the elemental constituents within biological tissues down to trace levels of concentration. The knowledge obtained this way is very important for gaining new insights of the highly complex functions of the elements within the living tissues and cells. While light microscopy and electron microscopy provide particularly useful information about the composition and structure (shape) of tissues and cells, the information of the elemental composition of the tissues from major to trace levels, and their chemical states obtained by synchrotron radiation-based spectroscopy can complement the structural information. As a simple demonstration of the techniques and topics discussed in this book, five images of a cell (a section or the whole cell) are shown in Fig. 1.1. These images are:

(a) The light microscopic image of a single cell.
(b) The scanning electron microscopic image of a single cell.
(c) The elemental distribution within a single cell.
(d) The transmission electron microscopic image of a section of the cell.

(e) The transmission electron microscopic image of the elemental distribution within a section of a single cell.

The figures clearly show that we can obtain very different information that can be related to different functions of a single cell.

Synchrotron radiation is an electromagnetic radiation that is emitted by charged particles moving at relativistic speeds in circular orbits in a magnetic field. Synchrotron radiation was initially considered a nuisance in high-energy accelerators and particle storage rings, electron accelerators in particular. The energy loss as a consequence of synchrotron radiations is especially significant in electron synchrotrons and storage rings. In the 1950s and 1960s, synchrotron radiation was eventually discovered as a very useful X-ray source that can be used in numerous fields in science. Many accelerator laboratories around the world started research projects for constructing devoted storage rings to make use of the radiations. The radiation used for these purposes is primarily in the ultraviolet and X-ray wavelengths, characterized by high brilliance and tunable energy.

The synchrotron radiation facilities have been increased in number and enlarged in size. European Synchrotron Radiation Facility (ESRF) in Grenoble (France) and SPring-8 (Super Photon Ring) in Harima (Japan) are the two largest facilities that became operational in the last decade of the twentieth century. As of 2003, there are about 50 radiation facilities in operation or under construction around the world, 17 in Europe, 12 in Japan, 10 in United States, and the rest in the other countries around the world. These facilities are operated serving user communities from universities, research institutes and industries. Very exciting research topics in the fields of material science, biology, and medicine are currently under investigation in these facilities.

The diversity of the techniques and applications of synchrotron radiation makes it impossible to be covered in a single monograph. In this book, selected principles and applications in certain fields of biology and medicine will be presented. To make the book readable for the potential researcher in biology and medicine, the physics and technical principles are not discussed in details. The author believes that there are very good books and papers covering the physics and technological aspects of synchrotrons and synchrotron radiation based on the wealth of X-ray physics for over one century.

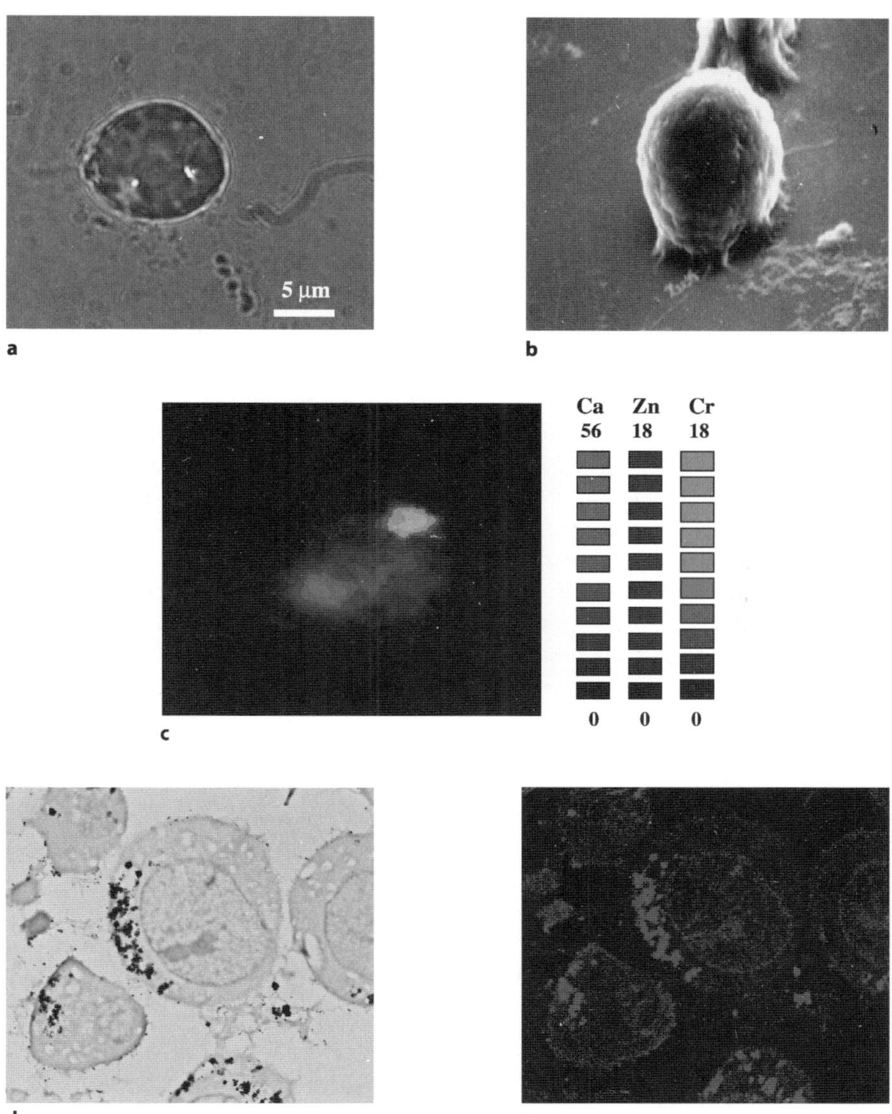

Fig. 1.1a–e. Images of a single cell using different techniques: (**a**) light microscopic image of a single cell, (**b**) scanning electron microscopic image of a single cell, (**c**) elemental distribution within a single cell, (**d**) transmission electron microscopic image of a section of a single cell and (**e**) transmission electron microscopic image of the elemental distribution within a section of a single cell

2

Synchrotron Radiation and X-ray Fluorescence Spectroscopy

2.1 Synchrotron Radiation

Synchrotron radiation (SR) was observed for the first time in April 1947 at General Electric in an advanced type of accelerator, an electron synchrotron [1]. While initially it was considered a nuisance, in the 1950s it became clear that the source of energy loss and annoyance for accelerator designers might become a very useful source of X-rays with potential applications in material science [2].

When electrons or positrons moving at relativistic speed, i.e., close to the velocity of light, are subjected to a magnetic field, the trajectory follows a circular orbit and the SR is emitted in the tangential direction. The energy of the SR covers a broad spectrum with a peak at the so-called critical energy E_c, which is proportional to the electron energy E and inversely proportional to the radius of the trajectory ρ, according to:

$$E_c(\text{keV}) = 2.218\, E^3\,(\text{GeV})/\rho(m)\,. \tag{2.1}$$

The rate of emission of the photon, or the photon flux F (photons/s/0.1% beam width), the flux density D (photons/s/mrad^2/0.1% beam width) and the brilliance B (photons/s/mm^2/mrad^2/0.1% beam width) of the synchrotron radiation from electrons in a circular orbit in a constant magnetic field is given by:

$$F = 8.67 \times 10^9 I(mA)\gamma\,(E/E_c)^{3/2}\,K_{2/3}^2\,(E/2E_c)\,\Delta\theta$$

$$D = F\Big/\left(2\pi\sqrt{\sigma_{x'}^2 + \sigma_{p'}^2}\sqrt{\sigma_{y'}^2 + \sigma_{p'}^2}\right) \tag{2.2}$$

$$B = D\Big/\left(2\pi\sqrt{\sigma_x^2 + \sigma_p^2}\sqrt{\sigma_y^2 + \sigma_p^2}\right) \tag{2.3}$$

where:

I is the electron beam current
$\gamma = 1957 \times E$ (GeV) is the relativity parameter

$K_{2/3}$ is the modified Bessel function of the second kind

σ_x, σ_y are the horizontal and vertical size of the electron beam

$\sigma_{x'}$, $\sigma_{y'}$ are the horizontal and vertical beam size of the photon beam

σ_p, $\sigma_{p'}$ are the divergence of the electron and of the photon beams, respectively.

As an example, the brilliance of the SR generated in a bending magnet in SPring-8 is shown in Fig. 2.1.

What makes X-rays from a synchrotron radiation source so useful leading to their wide use in physical, chemical and biological field are their unique properties:

1. Tunability in incident X-ray energy. The absorption coefficient of an element is influenced by the chemical states of the substances, giving rise to the analytical method XAFS-X-ray absorption fine structure spectroscopy. Using this feature, one can obtain the information of chemical state of elements.
2. High photon flux. The emitted radiation has high intensity, 10,000 times higher than conventional X-ray tubes. This feature results in high efficacy by reducing the measurement time.
3. High collimation. The highly collimated SR is suitable for microanalysis. This feature enables analysis of trace metallic elements contained in a biological specimen at a single cell level.

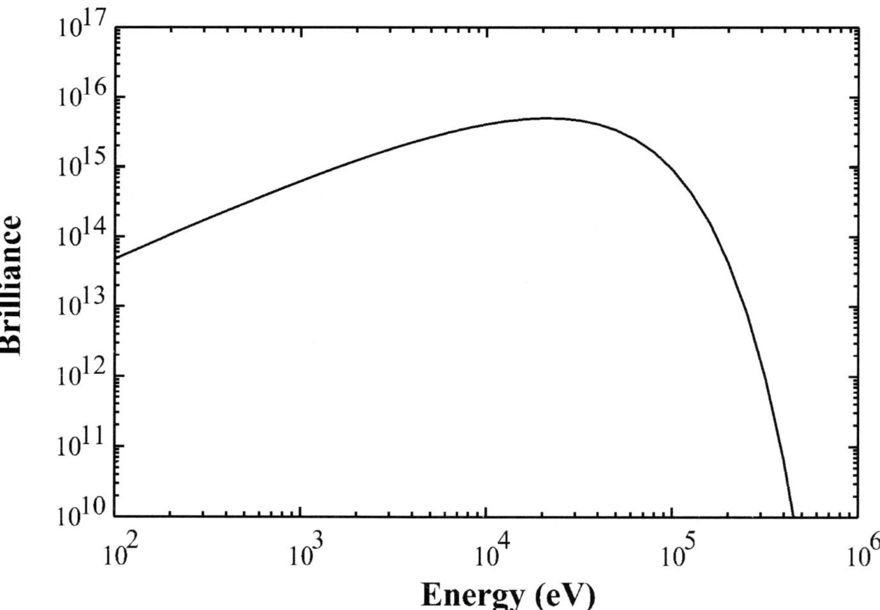

Fig. 2.1. Calculated brilliance versus photon energy for the bending magnet in SPring-8. The unit of brilliance is photons/sec/mm^2/μrad^2/0.1% beam width

4. Pulsed time structure. Photons radiated from bunched electrons running periodically in the storage ring are pulsed at controlled intervals. This feature makes it possible to perform time-resolved measurements.
5. The radiation has a selective distinct linear or circular polarization.

These features make the SR, in many cases, the only means of localized, non-destructive analyses of materials with extremely low concentration and thus most suitable for biological samples application. The contents of this chapter are mainly focused on one aspect of application of synchrotron radiation, namely X-ray fluorescence spectrometry, quantification and energy selective fluorescence spectrometry.

2.2 Advances in Synchrotron Development

Early in SR history, SR research was performed in a "parasitic" facility of a high current accelerator laboratory for high-energy or nuclear physics (the first generation). The 1980s saw the design and construction of dedicated SR facilities, the second generation. And in 1990s the third generation facilities were developed, using optimized magnet lattice and insertion devices in order to obtain more beam brilliance than bending magnets. The third generation SR facilities can generate $10^{11\sim12}$ times higher brilliance than laboratory X-ray tubes.

SR facilities typically consist of an injection system, a storage ring and beam lines. In the injection system, electrons are generated, pre-accelerated, and sometimes a second accelerator further accelerates these electrons to more than 1 GeV before injection into the storage ring.

In the ring, bunches of electrons periodically circulate at relativistic speed for periods of up to many hours. The storage ring consists of radio-frequency (RF) cavities, bending magnets, other magnets, insertion devices and other control systems (Fig. 2.2).

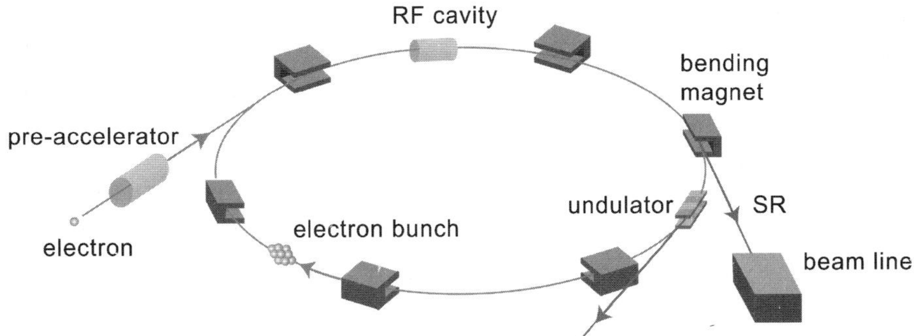

Fig. 2.2. Schematic of the electron storage ring of a SR facility

RF cavities. The RF cavity system restores energy, which the electrons lose because of the emission of SR, and stabilizes the bunch of electrons by phase-stability principle. The frequency of acceleration voltage is fixed to an integral multiple of the orbital frequency. That is, in the phase of RF, the voltage is synchronized when electron come to the RF cavity on the reference orbit (Fig. 2.3). Electrons that are slightly fast, get less acceleration and slow down because the phase of the acceleration voltage is ahead and the RF voltage is smaller. On the other hand, electrons that are slightly slow get more acceleration and speed up. Thus, the electrons exhibit longitudinal oscillations around the reference center of the bunch (called synchrotron radiation), and the bunch of electrons that are accelerated together is stabilized.

Bending magnets. Bending magnets bend the trajectory of electrons and force them to circulate in orbit. Synchrotron radiation is emitted when an electron received centripetal force in the magnetic field of the bending magnet. Synchrotron radiation emitted from an electron traveling at almost the speed of light is highly collimated by relativity effect. The magnitude of the relativistic angular width of the bending magnet radiation ($\Delta\psi$) is given by:

$$\Delta\psi = 1/\gamma\,. \tag{2.4}$$

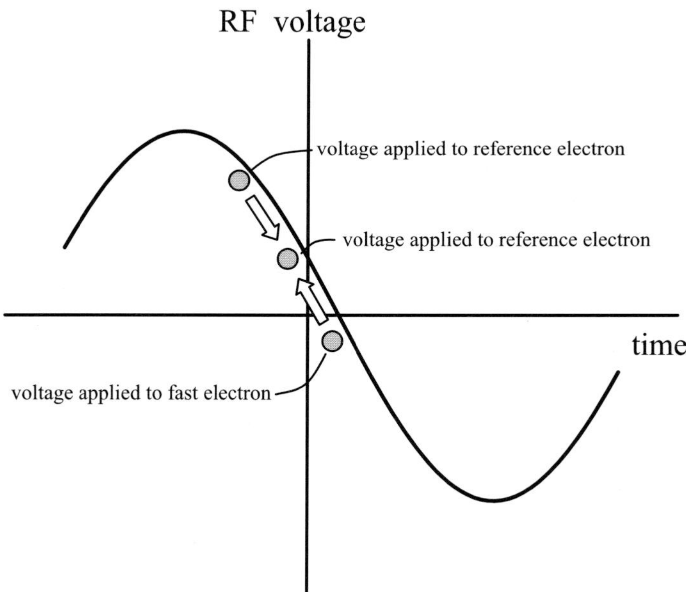

Fig. 2.3. Plot of radiofrequency cavity voltage versus time. The cavity voltage at the time zero shows the voltage which is seen by the reference electron passing through the cavity

The critical wavelength is given by:

$$\lambda(c) = 18.64/(B * E^2) \tag{2.5}$$

where B (the magnetic field) is in Tesla and E (the beam energy) is in GeV. One-half of the power is radiated above and one-half below the critical wavelength.

In the case of SPring-8, which has an 8 GeV storage ring, the angular radiation is about 60 micro-radians, corresponding to 3 mm beam size at 50 m from the source.

Insertion devices. Higher intensity synchrotron radiation is produced by an insertion device. The insertion device is comprised of a periodic array of dipole magnets with alternating polarity. According to the magnitude of the oscillation of the electron trajectory, there are two types of insertion devices, an undulator and a wiggler. The insertion device is installed in a straight section of the electron trajectory in the storage ring. As electrons pass through the insertion device, the trajectories of electrons wiggle several times and the electrons emit synchrotron radiation. The radiation cones emitted at each bend in the trajectory give rise to interference effect.

In a **wiggler**, a sequence of bending magnets with relatively weak magnetic field results in a small deflection angle, $(< \gamma^{-1})$, and the interference effects produce a radiation which has a continuous spectrum with higher fluxes and with short wavelengths. The wiggler is often used as a source in order to increase the flux at shorter wavelengths. A sequence of bending magnets with n poles of alternating polarities can enhance the flux by $2n$ times (the upper smooth curve of Fig. 2.4). The critical wavelength for a wiggler is lower than that of a bending magnet.

In an **undulator**, a periodic array of strong magnets resulting in a large deflection angle $(\gg \gamma^{-1})$, and the coherent interference effects produce highly collimated radiation, which has one or a few spectrally narrow peaks (a fundamental one and harmonics). For n poles, the beam's opening angle is decreased by $n^{1/2}$ and thus the intensity per solid angle increases as n^2 (the upper curves of Fig. 2.4).

In the case of a helical undulator at beam line 40XU in SPring-8, for example, the energy of the fundamental radiation is concentrated in the core. The angular spread of the central radiation is only 15 µrad (horizontal) × 5 µrad (vertical), corresponding to 0.75×0.25 mm at 50 m from the source, and the flux is as high as 1.5×10^{15} photons/s.

A comparison of the brilliance of the SR from different sources (Fig. 2.4) shows that the synchrotron radiation from a bending magnet is about eight orders of magnitude higher than that of conventional X-ray tubes and in the energy range of 1–100 keV, which is further enhanced by an order of magnitude of two by a wiggler and four by an undulator.

The synchrotron radiation beams are usually fed into experimental areas through slits, focusing mirrors, and monochromators.

A **monochromator** is used to select a very narrow energy band of the spectrum and a focusing system (e.g., Kirkpatrick–Baez mirror system) can be used to obtain submicron beam diameter. A synchrotron radiation beam can thus be within a few microns in size and can have a variable (tunable) energy.

The end statin consists of instruments for introducing samples to the beam and associated detectors for measuring the original and fluorescent ra-

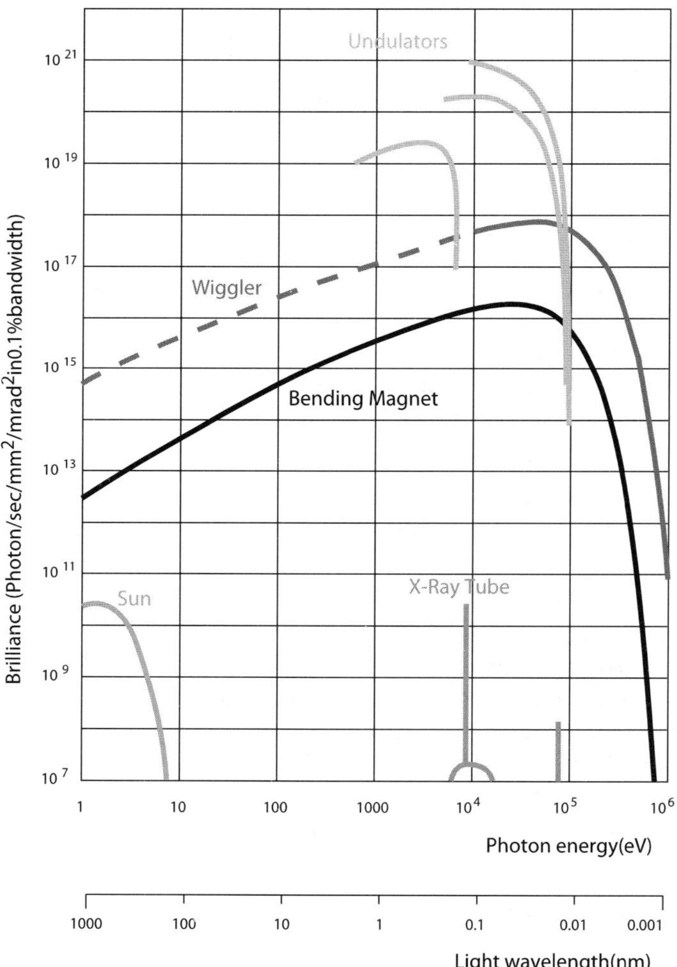

Fig. 2.4. The comparison between the light from synchrotron radiation and that from conventional X-ray sources. It can be seen that synchrotron radiation X-ray is about one billion times more brilliant than conventional X-ray sources. This figure was quoted from reference [15]

diations, including instrumentation to control the end station and measuring the response of the detectors.

With the availability of hard X-ray radiation even large pieces (several cm^3) can be investigated. By tuning the energy of the X-rays, one can further adjust the penetration depth, thus achieving surface or bulk sensitivity. Varying the incidence to grazing angles, the X-ray beam can be used as a surface probe. Because of the very small divergence, it enables investigation of thin layers or coatings. The high intensity allows very fast data acquisition, which in turn enables measurements to be carried out during processing of the material under investigation. These in situ experiments provide information about the dynamics of processes, which take place during the transformation of the sample. The large penetration depth of X-rays facilitate the study of samples in different sample environments such as furnaces, cryostats, pressure or chemical cells.

2.3 Examples of Synchrotron Radiation Facilities

Most of the experimental results that will be presented in this book were obtained at two synchrotron radiation facilities in Japan: the Photon Factory (second generation) and SPring-8 (third generation). The characteristics of the SR beam and beam lines of these facilities are representative of SR facilities worldwide and are described here.

1. *Photon Factory (PF)*

   Photon Factory (PF), High Energy Accelerator Research Organization (KEK), a second-generation facility, started experiments in 1983. In PF, electron energy in the storage ring is 2.5 GeV and maximum current is 400 mA. The parameters of the storage ring are summarized in Table 2.1.

   – *Experimental layout of BL4A*

     The layout of beam line 4A is shown schematically in Fig. 2.5. Synchrotron radiation from the storage ring is monochromatized with a multilayer monochromator. Incident X-rays are focused using Kirkpatrick–Baez optics [3]. The beam size incident on the sample is about $6 \times 5\,\mu m^2$. The incident and transmitted photon flux is monitored with ionization chambers. The incident photon flux is

Table 2.1. The parameters of the storage rings in Photon Factory and SPring-8

	Photon Factory	SPring-8
Beam energy	2.5 GeV	8 GeV
Initial current beam	400 mA	100 mA
Circumference	187 m	1,436 m
Emittance	6.8 nm	36 nm

$10^8 \sim 10^{10}$ photons/s. The fluorescent X-rays are collected by a solid state detector (SSD), with a 0.3 mm thick Be window, 12 mm^2 active area, and 160 eV resolution at 5.9 KeV. The angle between the incident beam and the detector is fixed at 90°. The sample microstage has an $x - y$ motion on a vertical plane against the beam, and has $\theta - 2\theta$ motion around the horizontal rotation axis, driven by stepping motors. The sample can be viewed at high magnification (60×) using a CCD camera during measurement.

2. *SPring-8*

SPring-8, Japan Synchrotron Radiation Research Institute (JASRI), a third-generation synchrotron radiation facility, opened for research in 1997. The SR facilities consist of linear accelerator, storage ring, beam lines, etc. One of the most important parts determining the characteristics of SR is the electron or positron storage ring. In SPring-8, the electron energy in the 1.5 km circumference storage ring is 8 GeV, and the maximum current is 100 mA. The parameters of the storage ring are summarized in Table 2.1. SPring-8 is equipped with advanced, high-performance insertion devices, resulting in high brilliance as high as 10^{20} photons/s/mm^2/mrad^2 in 0.1% beam width. These high fluxes of photons make it possible to analyze ultra- trace elements contained in a small area in biological tissues.

– *Experimental layout of BL39XU*

BL39XU is a hard X-ray undulator beam line that is mainly used for studies of ultra-trace element analysis. A combination of fundamental and third harmonics of the undulator radiation covers an energy range from 5 to 37 keV. The layout of beam line 39XU is shown schematically in Fig. 2.5. Synchrotron radiation from the undulator radiation is monochromatized with a Si(111) double crystal monochromator. The third harmonics from the undulator radiation is cut off to less than 10^{-4} by a platinum-coated mirror of horizontal deflection, if needed. XRF and XAFS analysis for ultra-trace elements can be carried out in this beam line. Incident and transmitted photon flux is monitored by air-filled ionization chambers. Incident photon beams are restricted by a set of $x - y$ slits and a pinhole. Incident beam size is about 10 μm in diameter. Excited fluorescent X-rays are detected by a Si(Li) solid state detector. The angle between the incident beam and the detector is fixed to 90°, where the scattered radiation is minimized due to the polarization of SR [4]. Measurements are performed in a sample chamber and in vacuum.

Fig. 2.5a–c. Schematics of the beam line layout and experimental setup for X-ray fluorescence analysis and X-ray absorption fine structure analysis. (**a**) Photon Factory beam line 4A, (**b**) SPring-8 beam line 39XU, (**c**) SPring-8 beam line 40XU

a PF BL4A

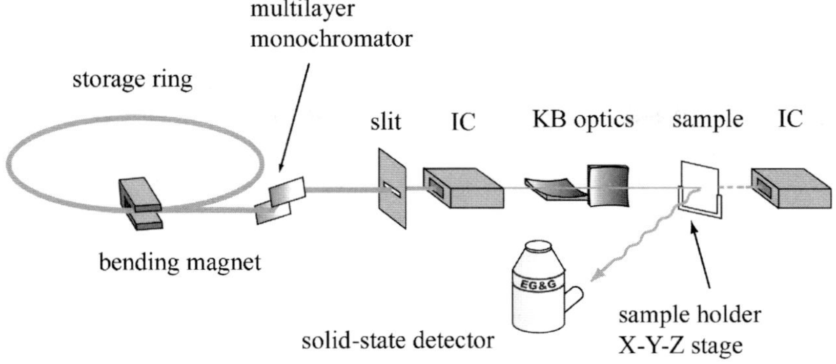

b SPring-8 BL39XU

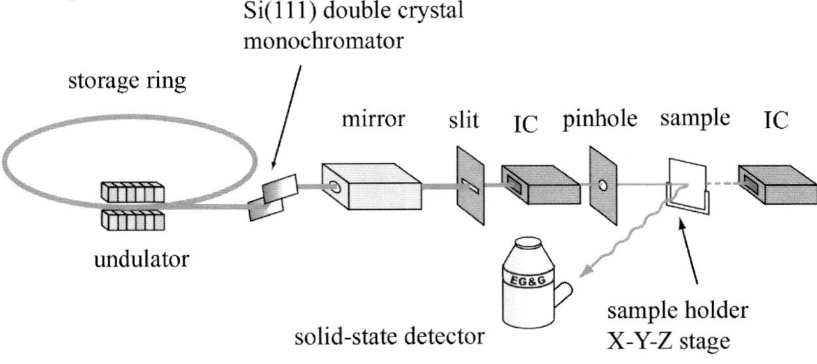

c SPring-8 BL40XU

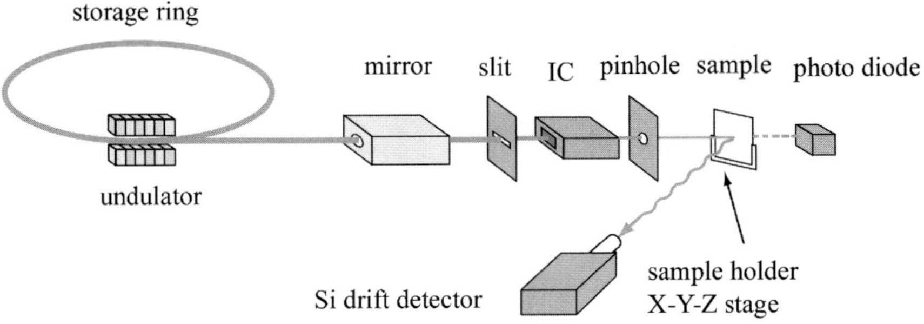

– *Experimental layout of BL40XU*

BL40XU is a helical undulator beam line with a high photon flux, without using a crystal monochromator. The core of its radiation has an energy spectrum with a very sharp fundamental peak (energy peak-width of 2%), and thus the flux is more than 100 times higher than that obtained with a crystal monochromator. Even when only the central 15 μrad (horizontal) × 5 μrad (vertical) radiation is used, the flux is as high as 1.5×10^{15} photons/s. The fundamental radiation covers an energy range from 8 to 17 keV by varying the undulator gap.

Using the characteristics of high photon flux, one can perform fast XRF imaging in this beam line. Incident X-rays are restricted by a pinhole made of tantalum (φ 2.4 μm). Transmitted X-rays are monitored by a PIN photodiode. Fluorescent X-rays are collected by a Si(Li) detector (max 30,000 cps). The sample stage has an $x - y$ motion on vertical plane against the beam, driven by stepping motors. A CCD camera is used to monitor the sample during a measurement.

2.4 Synchrotron Radiation X-ray Fluorescence Analysis (SR-XRF)

2.4.1 Fluorescence X-ray

The synchrotron radiation X-ray source represents a great improvement in X-ray production technology [5]. By using synchrotron radiation for XRF analysis, many researchers can make measurements with improved sensitivity and spatial resolution. The major features of SR-XRF analysis are as follows:

1. local area analysis by using microbeams
2. possible to measure in the air or water
3. non-contact and non-destructive assay
4. rapid measurement
5. precise assay for the trace elements.

Especially, the features of 2., 3. and 5. are superior to the other elemental analysis techniques.

By means of μ-XRF analysis using synchrotron sources, one can collect information on the distribution of trace constituents of a material with high lateral resolution. In view of the high sensitivity for heavy elements, synchrotron radiation induced μ-XRF is particularly valuable for the trace-level microanalysis of the heterogeneous geological materials and the biomedical samples [6].

The generation of the fluorescence X-ray is caused by the excitation of an inner shell electron and the transition of another electron from an outer shell to this vacancy. Energy release takes place either by emission of an

Auger electron or by a quantum of the so-called characteristic X-ray. The de-excitation of an atomic shell is characterized by the **fluorescence yield**, which is defined as the number of characteristic X-rays per primary vacancy emitted from this shell. The corresponding **Auger yield** is defined as the number of Auger electrons emitted per primary vacancy from the shell.

For atomic shells other than the K shell, each individual sub-shell has both a fluorescence yield and Auger yield, and the flux of X-rays is related to the initial distribution of vacancies among the sub-shells. The phenomena become more complicated by the occurrence of so-called Coster–Kronig transitions in which the initial vacancy is transferred from one sub-shell to another. For the K shell, the situation is relatively simple because only two values, ω_k and $a_{k,}$, defined as the K shell fluorescence and Auger yield, respectively, are involved. These are connected by one relation ($a_k + \omega_k = 1$), and have been subjected to many experimental and theoretical studies. However, for the L shell, instead of two yields, there are nine yields connected by three relations [7]. When the atomic number Z is high, the value of the fluorescence X-ray yield becomes higher. If the atomic number Z is not too small ($Z \geq 11$ for the K series and $Z \geq 30$ for the L series), the characteristic radiation lines lie in the measurable X-ray energy range (above 1 keV) [8].

2.4.2 Detectors

The fluorescence X-rays emitted by an atom after absorbing the synchrotron radiation are measured in detectors for quantitative analysis. The most commonly used X-ray detectors are the proportional counters and solid-state ionization detectors, namely scintillation detectors, Si(Li) detectors and charge coupled devices. The detectors measure the X-rays energy spectrum, specifically the characteristic X-rays, whose intensities are proportional to the concentration of the elements in the sample. Solid state detectors are energy dispersive detectors, where the X-rays are absorbed and converted to electrical signals proportional to their energy. These detectors usually have good energy resolution, defined as that required to resolve the different characteristic X-rays. The proportional counters can be operated in an energy dispersive mode, but the energy resolution is usually not adequate. In order to improve the resolution, the proportional counter is used as a detector only in a Bragg crystal spectrometer, where the energy dispersion is carried out by Bragg filters. This detector system is called the wavelength dispersive spectrometer and is particularly suitable for lower energy X-rays (< 20 keV).

Solid-state detectors cover a wide energy range (1–100 keV and higher) and provide a large solid angle of detection. The latter is very important in analysis of extremely small amounts of materials. For energies higher than 20 keV, wavelength-dispersive systems are not very effective and the SSD is the most convenient device to detect the K lines of heavy elements [8].

A solid state detector is essentially a reverse-biased diode with a wide junction (to a few mm) of carrier free, depleted layer created by compensating p-type material with n-type donors such as Li. Li drifted Si detector, Si(Li) for short, is the most commonly used type, and lately inherently intrinsic Ge detector is also used for X-ray spectrometers. In these detectors, ionization by the X-rays in the intrinsic region produces electron-hole pairs that are swept by the reverse bias electric field to produce a current pulse. The number of pairs created is proportional to the incident X-ray energy. The charge collected at the anode is converted to a voltage by an amplifier. These are subsequently converted into voltage pulses by a preamplifier and are further amplified and shaped by a linear amplifier to optimize the signal-to-noise ratio. The signals are then fed into a multi-channel pulse height analyzer to be sorted into an energy spectrum. Solid state detectors are stored in liquid nitrogen to prevent the diffusion of lithium of the depletion layer in the Si(Li) case and to reduce the noise in general.

In the simplest form of data acquisition, a certain range of energy corresponding to the characteristic X-ray for an element can be selected using a single channel pulse height analyzer (SCA). X-ray line scans and X-ray maps for the element can be obtained by recording the intensity of this energy window as a function of the sample coordinates.

This simple procedure is very useful for on- or off-line explorative data analysis, but implicitly assumes that within the energy window used, a single, non-overlapped peak is present with a high peak-to-background ratio so that the integrated intensity within the window is a good estimate of the net intensity of the peak. Unfortunately, for XRF spectra in general, these assumptions are not valid; peak overlap frequently occurs in energy dispersive X-ray spectra especially for peaks corresponding to trace elements [6]. It is necessary to estimate the integrated value of the overlapped area and background in order to realize the precise quantification.

2.4.3 X-ray Fluorescence Spectrometry: A Typical Spectrum

A typical X-ray fluorescence spectrum is shown in Fig. 2.6. The y axis (ordinate) shows the intensity (total counts per channel in the MCA) as a function of the energy (x axis, abscissa) of the detected X-ray. The sample measured here is a very thin specimen (about $6\,\mu m$) of a brain tissue. The X-ray beam is $10\,\mu m$ in diameter. The relevant elements are marked on each peak, corresponding to the characteristic X-ray peaks of that element. A table of the characteristic X-ray energy of the elements (K_α) is cited by J.W. Mayer and E. Rimini [9]. The height of the peaks, more precisely the area under a peak, is proportional to the concentration of the element in the specimen. A careful examination of the spectrum shows that the peaks representing each element are superimposed on a "back ground" signal.

The spectrum usually contains a number of spurious discreet components, appearing as peaks, namely the sum peaks (also known as pile-up peaks) and the escape peaks.

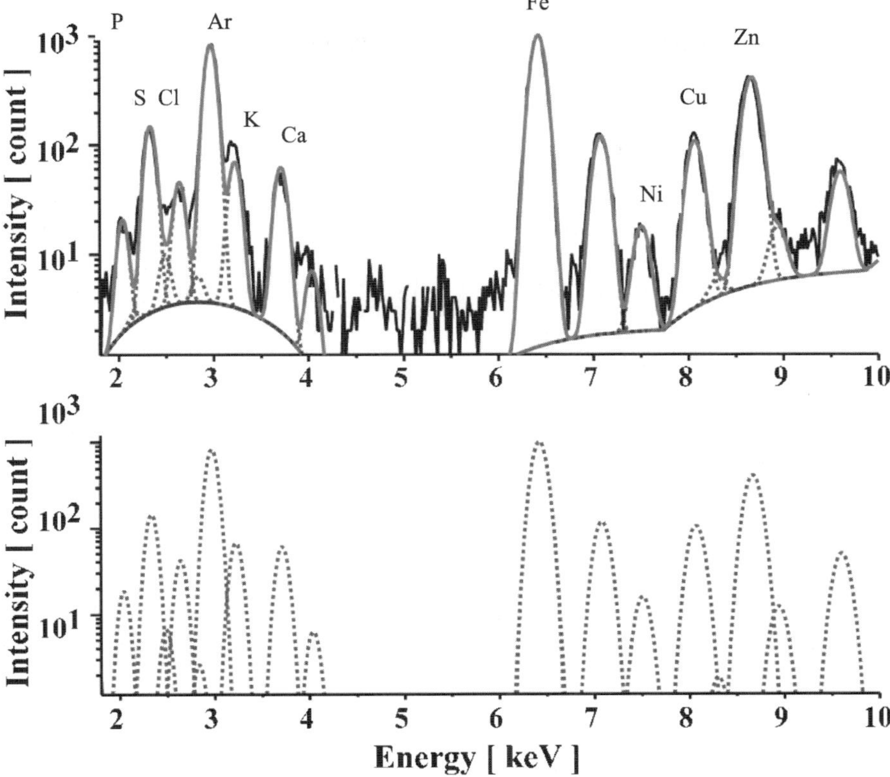

Fig. 2.6. Typical X-ray fluorescence spectra before and after peak separation and background reduction

Sum peak: A peak that appears at an energy that corresponds to the sum of two or more other peaks' energies. A sum peak occurs from the summing of the electrical pulses at high-count rates because the individual nuclear or electronics events occur within a time period that is less than the resolving time of the amplifier. Therefore one count is lost from each peak and is added to the sum peak.

Escape peak: A peak displaced at the lower energy side by a well-defined amount. The escape peak arises from the escape of the K X-ray of detector material. In the case of Si detectors the escape peaks are expected to appear at the channel corresponding to energy $(E - 1.74)\,\text{keV}$. The escape peak is typically less than 1% of the peak of the element of interest.

2.4.4 Background Level in Detected Spectrum

2.4.4.1 Basic Components of the Background in SR-XRF

The interaction between the exciting radiation source and a substance has complicated properties because of a variety of components added to the sample emission spectrum. When the incident X-ray beam has a large flux of quanta, the complicated factors are the spectral density of the background in the region of the lines being analyzed and the counting rate of the detector that mainly determine the analytical sensitivity. In the most common case of excitation by monochromatic radiation of samples with a light-element matrix, the background has the following forms:

1. a peak from the elastic scattering of the exciting radiation;
2. a peak from the Compton scattering of the same radiation (single for thin samples and multiple for thick ones);
3. escape peaks of detector fluorescence emission, spaced from the elastic and Compton peaks as well as from the rather intense spectral lines of the sample at the energy of the fluorescence quantum emission of the detector material (9.9 keV for Ge K_α);
4. escape peaks of detected quanta, Compton-scattered in the detector at an angle of about 180° and leaving it along the shortest path. In this case, the detector registers the recoil electron alone. When there are many intense lines in the spectrum of a sample or when the detector is of insufficient thickness, the indicated peaks form a continuous background in the low-energy region [8].

2.4.4.2 Compton Scattering

An electromagnetic wave has properties as both the wave and particle. In the energy region of X-ray, it tends to display the property as a particle. In the process of inelastic scattering by free or comparatively weakly coupled electrons in a substance, X-ray quanta lose some fraction of their energies. The incident X-ray is scattered as one with a little longer wavelength. This phenomenon is called Compton scattering. If the direction and detection solid angle are optically chosen, the application of the polarized synchrotron radiation enables the intensity of this process to be considerably reduced (by a factor of 10–100) and, thus, the sensitivity to be improved or the measurement time to be shortened using XRF technique.

2.4.4.3 Elastic Scattering

The electrons of a substance are forced to oscillate by incident X-ray. The subsequent emission of X-ray with the same frequency is caused by their oscillations. This phenomenon is called elastic scattering. The atomic nucleus of a substance is forced to oscillate by incident X-ray, too. However, it is generally neglected because of its weak amplitude. With excitation by the white or wide-band SR beam, the elastic scattering can become the basic background below the analytical peak of a spectrum if the cross section of the elastic scattering towards a SSD starts to exceed the Compton scattering. With monochromatic excitation the peak from the elastic scattering lies outside the detected emission lines of elements and can influence the analytical sensitivity, creating an additional load for a SSD and a spectrometric amplifier. In this case, the effect will be significant if the elastic-scattering cross section becomes larger than or comparable to the Compton-scattering cross section [8].

2.4.4.4 Bremsstrahlung Radiation of Photoelectrons in the Sample

Under the interaction of the exciting radiation and a sample, a certain number of photoelectrons appear and are decelerated in the sample volume, as a result of the photoeffect. During their decelerations, there appears radiation with a continuous spectrum lying within the $E_\gamma \leq E_\varepsilon$ range of energies where E_ε is the energy of the photoelectrons. Among the major factors, it is suggested that limiting the ultimate sensitivity of the SR-XRF technique is the bremsstrahlung radiation background under the analytical peaks [8].

2.4.4.5 Improvement for Reducing the Background

It is suggested that the presence of the natural SR polarization is the most important qualitative advantage of SR for the X-ray fluorescence analysis process over the other types of X-ray radiation. The plots of the elastic and inelastic scattering of the exciting radiation have a minimum that depends on the mean polarization co-effect, the selected value of energy and the angular opening of collimators in the detection system. The decrease of the intensity of the elastically scattered and the Compton peaks can be realized by placing the detector in the plane of the E-vector of the monochromatic SR beam at an angle $\theta = 90°$ to the beam and consequently improve the background plateau height of an incomplete charge collection in the SSD. In addition, the acceptable rate of counting the fluorescent lines grows because of the limited counting rate of a SSD.

If the emission spectrum has peaks in many orders of magnitude different in intensity, the excitation of some parts of a spectrum would better be done separately by varying the excitation conditions (monochromatic energy is set higher and lower than the K absorption edge of an element of high concentration). The selective excitation makes possible to suppress the heavy-element lines during detection of fluorescent quanta of lighter elements.

The use of the specific features of SR makes possible to reduce substantially the background below the analytical peaks and, hence, to improve the sensitivity and to shorten the time of analysis [8].

2.5 Quantitative XRF Analysis

2.5.1 Basic Equations

The schematic representation of the interaction between the incident X-ray and the sample is shown in Fig. 2.7. The incident X-rays with a wavelength λ, which irradiates the sample, are attenuated and generate scattered X-rays (Rayleigh scattering and Compton scattering) and fluorescent X-rays, accompanied by the generation of photoelectrons. When the sample is a pure element i, according to the Beer–Lambert law, the intensity of the transmitted X-rays with wavelength λ, $I(\lambda)$, is given by

$$I(\lambda) = I_0(\lambda)\exp\{-\mu_i(\lambda)\rho_i t\} \tag{2.6}$$

where the $I_0(\lambda)$ is the intensity of the incident X-ray with wavelength λ, $\mu_i(\lambda)$ is the effective mass attenuation coefficient of element i for incident wavelength λ, ρ_i is the density of element i, and t is the thickness of the sample. The total effective mass attenuation coefficient $\mu(\lambda)$ of a multi-element specimen is given by the simple relationship

$$\mu(\lambda) = \mu_1(\lambda)W_1 + \cdots + \mu_n(\lambda)W_n \tag{2.7}$$

where $W_1, \ldots, W_n$ are the weight fraction of components.

In the experimental configuration shown in Fig. 2.8, the intensity that reaches the depth of z, represented by $I(\lambda, z)$ is given by

$$I(\lambda, z) = I_0(\lambda)\exp\left\{-\mu(\lambda)\rho\frac{z}{\sin\phi}\right\} \tag{2.8}$$

where ϕ is the incidence angle and ρ is the density of the sample. The intensity of primary fluorescent X-ray generated in the sample between the depth of z and $z + \Delta z$, represented $\Delta I_\mathrm{f}(\lambda, z)$ is given by

$$\begin{aligned}
\Delta I_\mathrm{f}(\lambda, z) &= P_i\mu_i(\lambda)W_i\rho\left\{I(\lambda, z) - I(\lambda, z + \Delta z)\right\} \\
&= P_i I(\lambda, z)\left\{\mu_i(\lambda)W_i\rho\frac{\Delta z}{\sin\phi}\right\}
\end{aligned} \tag{2.9}$$

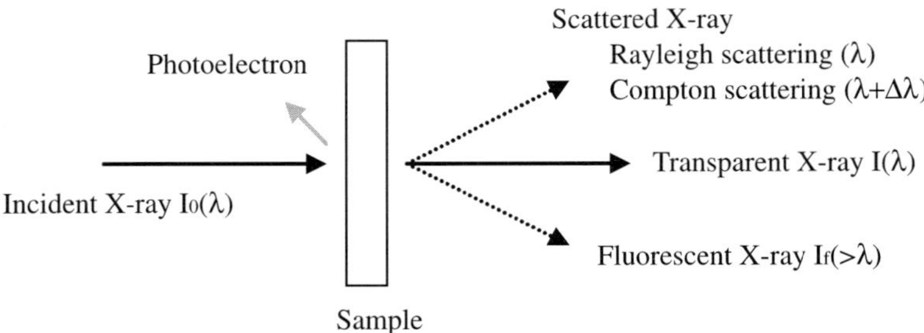

Fig. 2.7. The schematic drawing of the interaction between X-ray and the materials

where P_i is the probability that a characteristics X-ray line of element i is emitted, which is determined mainly by the ionization cross section and the fluorescence yield. The intensity of the primary fluorescent X-ray at the surface of the sample $\Delta I_\mathrm{f}(\lambda)$ is given by

$$
\Delta I_\mathrm{f}(\lambda) = \Delta I_\mathrm{f}(\lambda, z) \times \exp\left\{ -\mu(\lambda_\mathrm{f})\,\rho\,\frac{z}{\sin\varphi} \right\}
$$

$$
= P_i\, I_0(\lambda)\, \mu_i(\lambda)\, W_i\, \rho\, \frac{\Delta z}{\sin\phi}\, \exp(-\bar{\mu}\rho\, z) \tag{2.10}
$$

where

$$
\bar{\mu} = \frac{\mu(\lambda)}{\sin\phi} + \frac{\mu_\mathrm{f}(\lambda_\mathrm{f})}{\sin\varphi}\,, \tag{2.11}
$$

λ_f is the wavelength of the fluorescence X-ray, and φ is the take-off angle of the fluorescent X-ray detector. Therefore the total primary X-ray fluorescence yield I_f from a sample of a thickness t is given by

$$
I_\mathrm{f}(\lambda) = \int_0^t \Delta I_\mathrm{f}(\lambda)\mathrm{d}z
$$

$$
= P_i\, I_0(\lambda)\, \mu_i(\lambda)\, W_i\, \frac{1}{\sin\phi}\, \{1 - \exp(-\bar{\mu}\rho\, t)\}\, \frac{1}{\bar{\mu}}\,. \tag{2.12}
$$

In this study, the samples are thin sections and their major constituents are light elements. Therefore the magnitude of $\bar{\mu}\rho t$ small and the above equation can be approximated as

$$
I_\mathrm{f}(\lambda) = P_i\, I_0(\lambda)\, \mu_i(\lambda)\, W_i\, \rho\frac{t}{\sin\phi} \propto W_i\, \rho\, t\,. \tag{2.13}
$$

Under this approximation, the fluorescent X-ray intensity from each element is proportional to the incident X-ray intensity and to the area density

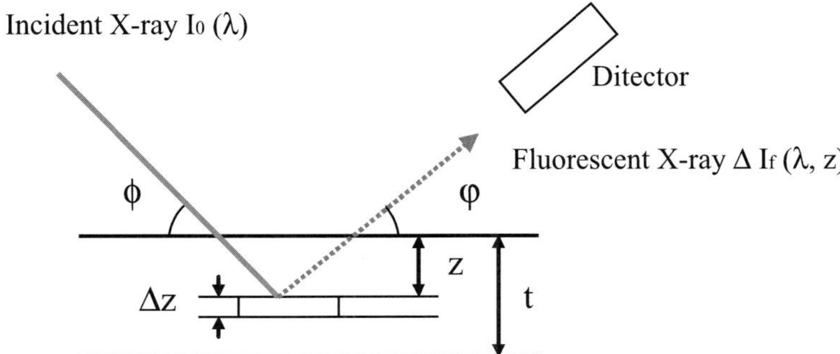

Fig. 2.8. The schematic drawing of the experimental configuration of the incident X-ray, the sample and the detector

of the element $W_i \rho t$. The fluorescent X-ray intensity that reaches the solid state detector $I'_f(\lambda)$ is given by

$$I'_f(\lambda) = C(\lambda)\, C_i\, I_f(\lambda) \tag{2.14}$$

where $C(\lambda)$ is the constant that is determined by the geometrical parameters of the set-up instruments, such as the solid angle to the detector and the path of the incident X-ray, and C_i is the constant determined by the attenuation coefficient of the path of the fluorescent X-ray about element i, which includes Be-window of the Si detector. $C(\lambda)$ and C_i are constants under the same experimental condition.

The fluorescent X-ray yield that reaches the detector is directly proportional to the peak areas $A_{\text{peak}\,i}$ of element i in the XRF spectra.

$$A_{\text{peak}\,i} = C' I'_f(\lambda) \tag{2.15}$$

where C' is the efficiency of the detector. Therefore the below equation can be derived from (2.13), (2.14) and (2.15).

$$A_{\text{peak}\,i} = \frac{1}{\sin\phi} C'\, C(\lambda)\, C_i\, P_i\, I_0(\lambda)\, \mu_i(\lambda)\, \rho'_i \tag{2.16}$$

where $\rho'_i = W_i\,\rho\, t = $ the area density of the element i.

C_i, P_i and $\mu_i(\lambda)$ can be obtained from the existing database and handbooks [9, 10]. $I_0(\lambda)$ was monitored by an ionization chamber. The product of C' and $C(\lambda)$ can be calculated by comparing the area density of element i and the peak area obtained from the reference sample, for which the area density of element i is already-known. It is, thus, possible to calculate the area densities of all elements from the XRF spectra because the product of C' and $C(\lambda)$ is independent of the elements. In this study, thin pure metal films whose thicknesses are known were used as the reference standards. After the determination of C' and $C(\lambda)$, the local area densities of all elements in the biological sample can be directly calculated from the measured XRF spectra.

2.5.2 Development of Computer Programs for Quantitative XRF Analysis

2.5.2.1 Objective

This chapter describes a computer code that has been developed for the quick processing of XRF spectra and quantification of the trace elements. This code can be used to investigate several important neurodegenerative diseases, such as Alzheimer's disease (AD), Parkinson's disease (PD), parkinsonism dementia complex (PDC), and amyotrophic lateral sclerosis (ALS), as well as to investigate basic biological samples in order to study changes in cells due to the incorporation of foreign metal elements.

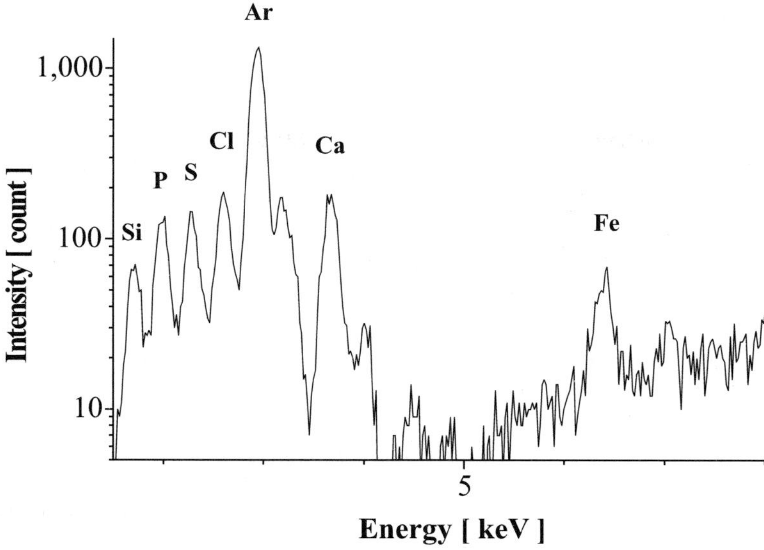

Fig. 2.9. The example of the XRF spectrum obtained from a biological sample. The fluorescent peaks from Si, P, S, Cl and Ar contents are strongly overlapping

The elemental area density can be calculated from the XRF spectra according to the equations described in the previous section. The typical constituents detected in the XRF analysis of biomedical samples are P, S, Cl, K and Ca, which are the main components of a living tissue. These peaks in the spectra overlap with each other making it difficult to decompose the peaks and calculate their areas accurately. The example of XRF spectrum is shown in Fig. 2.9, in which P, S, Cl and Ar peaks are strongly overlapping. In order to analyze such peaks appearing in spectra, many algorithms for background identification and peak discrimination have been reported in many fields for obtaining quantitative information, but various limitations exist tying each algorithm to specific spectral shapes [11]. The main purpose of these programs is to automate the numerical processing in obtaining correct quantitative data in the following three aspects [12]:

1. identification of background intensity
2. recognition of existing peaks
3. evaluation of positions and intensities

Few programs are available for the quantification using synchrotron radiation XRF. In this study, we originally designed the program that features the semi-automatic peak shape, energy and yield. It also includes graphical user interface and quantification procedures. In addition to the requirements described above, this program also has the following properties:

4. capability to analyze the spectra containing high noise

5. high flexibility to meet experimental conditions
6. simple and fast quantification from the peak areas
7. integration of multiple data

Biomedical samples generally contain heavy elements at low concentration, and the elemental distributions are not homogeneous. Therefore the spectra are not so clear as compared to those from minerals. High flexibility to the change of path length or experimental atmosphere is also important, because we use several different incident energies according to the experimental purpose, and some of the experiments are performed in air, others in vacuum. Features 6 and 7 are required to process large data obtained from large number of samples. The accumulation of the quantitative information is essential because the statistical accuracy is important in evaluating the biological function such as the differentiation of mouse embryonic stem cells or cell death in neurodegenerative disorders from the aspect of the elemental conditions. In these studies, it is necessary to understand the extent of residues and to obtain the average data as criteria. Recently the number of samples has increased due to the improvement of experimental efficiency and the increase of objective cases. The significance of systematic and fast processing of data has enhanced in accordance with the increase of samples. On the other hand, feature 2 is omitted in our program. In X-ray spectroscopy peak finding is usually not the crucial state of analysis because the peak locations are known beforehand [13].

2.5.2.2 Algorithm and Basic Equations for the Spectrum Analysis and Quantification

This program is written in Visual Basic, version 6. The quantification of trace elements is performed in the following procedures.

First, the analysis of the spectra obtained from reference samples are carried out. The reference samples are metal thin films whose thicknesses are already known. In this study, we analyzed pure films of Ti, Cr, Mn, Fe, Co, Ni and Cu. In this process, the peaks are fitted using the least squares method by varying the width, position and height of the peak after the background is estimated from the untreated spectra. The relations between the fluorescent intensity and the concentration are determined by comparing the peak areas from the samples with different thicknesses. The relation between channel and energy, the width and energy are also determined by comparing the position and the width of different kinds of the samples.

In the next process, the analysis of spectra obtained from autopsy specimens is performed based on the results from reference samples. In this process, only the heights of peaks have to be determined at the beginning because the width and the position have already been determined in the previous process. After the first coarse fitting by modulating the heights, the fine fitting is applied to each single peak in which the width, position and height of the peak are adjusted.

In the last process, the concentrations of the elements in the biomedical specimens are quantified by comparing the peak areas obtained in the second process to those of reference samples obtained in the first process.

The algorithm and the basic equation utilized in the program are described detailed in the following section.

Estimation of the Background

Quantitative analysis using XRF spectra requires the removal of the background prior to the estimation of the net area of the peaks. But the sample mass absorption coefficient, which is needed when calculating the background function, depends on the composition of the sample and is originally unknown. Therefore the extraction of the background is generally performed in several numerical ways [14]. In this program, the simple peak clipping approach was applied [15]. In order to process data from numerous spot analysis efficiently, a background approximation must be free of user-adjustable parameters to permit batch processing. The peak clipping approach provides rapid and robust estimation. This method is based on the equation represented by

$$Y_i = \min Y_i, m_i \qquad (m_i = Y_{i-1} + Y_{i+1}) \tag{2.17}$$

where Y_i is the count of channel i of the multi-channel analyzer, and m_i is the mean of the counts of channel $i - 1$ and $i + 1$. When the count of a certain channel i is compared to m_i, if Y_i is larger than m_i, it is replaced with m_i. This procedure is repeated for the selected extent of channels and the projections of spectra are gradually removed. The background of the spectra can be estimated easily by repeating this replacement. The pass count of 2,000 is chosen as the default value. Figure 2.10 shows the example of the estimation of the background in XRF spectrum. The solid and dotted lines show the untreated spectrum and the estimated background, respectively. The smoothing of the spectra can also be performed if needed.

Single Peak Fitting

Once the background has been evaluated, it is possible to carry out the peak fitting. The peaks are determined by the least squares method. Since the model functions for peak shapes are nonlinear with respect to their parameters, iteration is needed to perform the least squares fitting. The single peaks are fitted as Gaussian functions and are represented by

$$y(x) = H \exp\left\{ -\frac{(M - x)^2}{S} \right\} \tag{2.18}$$

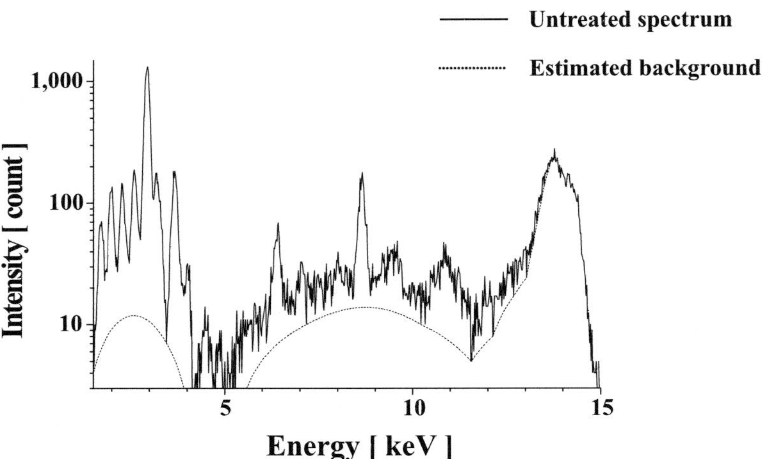

Fig. 2.10. Example of the estimation of the background in XRF spectrum obtained by the peak clipping approach

where M, S and H determine the position, FWHM (full width at half maximum) and height of the peaks, respectively, and $y(x)$ is the counts of the fitted peak in channel x. The objective function P is calculated via

$$P = \sum_i \{C(i) - B(i) - y(i)\}^2 \tag{2.19}$$

where $C(x)$ is the experimental counts obtained in channel x, $B(x)$ is the estimated background in channel x and i shows the extent of channels to which the fitting is applied. In this program, the steepest decent method was applied for a non-liner fitting of the peaks. This method modulates the variables of M, S and H as

$$M = M - \frac{\partial P}{\partial M}\,\mathrm{d}A, \quad S = S - \frac{\partial P}{\partial S}\,\mathrm{d}A, \quad H = H - \frac{\partial P}{\partial H}\,\mathrm{d}A \tag{2.20}$$

where $\mathrm{d}A$ is the step size. When these equations are directly applied, the absolute of the gradient P changes unstably, and M, S and H often diverge. Therefore, this program employs the modified steepest decent method, in which the variables are defined as

$$M = M - W_M \frac{\partial P}{\partial M}\,\mathrm{d}A, \; S = S - W_S \frac{\partial P}{\partial S}\,\mathrm{d}A, \; H = H - W_H \frac{\partial P}{\partial H}\,\mathrm{d}A \tag{2.21}$$

where W_M, W_S and W_H are the weight functions for M, S and H. These weight functions are determined empirically as

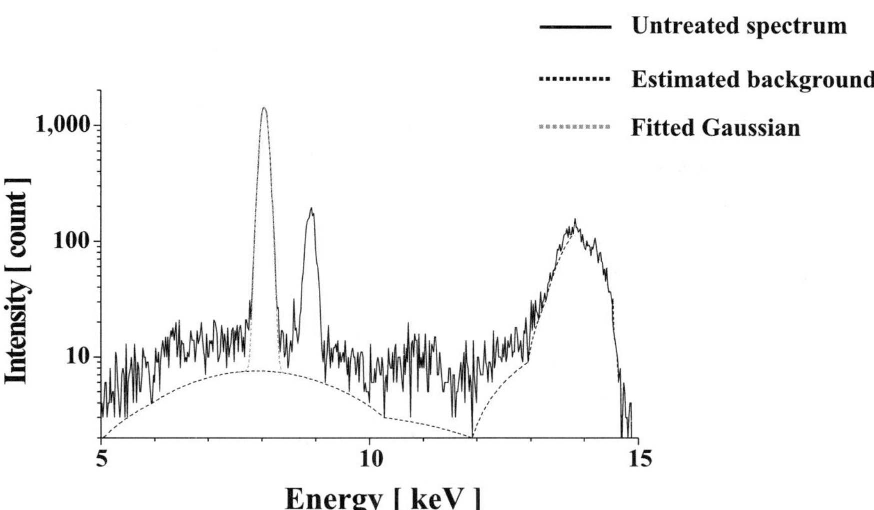

Fig. 2.11. Single peak fitting in the XRF spectrum, which was obtained from the reference sample of Cu thin film. It can be seen the fitted Gaussian is well correspondent to the measured spectrum

$$W_M = \frac{1}{H^2} \exp \frac{2}{S}, \quad W_S = \frac{20}{H}, \quad W_H = \frac{S}{5(1 - 4S^{2.5})}. \quad (2.22)$$

These weight functions are then normalized by the absolute value of the vector of (W_M, W_S, W_H) and used for the modulation of the variables. $dA = 1$ is chosen as the default number. The single peak fitting is performed by repeating the equation (2.21) until the objective purpose is minimized.

Figure 2.11 shows the example of the single peak fitting in the XRF spectrum, which was obtained from the reference sample of a Cu thin film. The black solid and dotted lines show the untreated spectrum and the estimated background respectively, and blue dotted line shows the peak obtained by the fitting. It can be seen the fitted Gaussian corresponds well to the measured spectrum.

Derivation of Calibration Curves

There are the relational expressions between the channel i of the multichannel analyzer and the X-ray energy E, and between the variable S that determines the FWHM of the peaks and X-ray energy E, which are represented by

$$i = aE + b \quad (2.23)$$
$$S^2 = cE + d \quad (2.24)$$

where a, b, c and d are the variables determined by the properties and configuration of the multi-channel analyzer and the detector [14]. Figure 2.12a,b

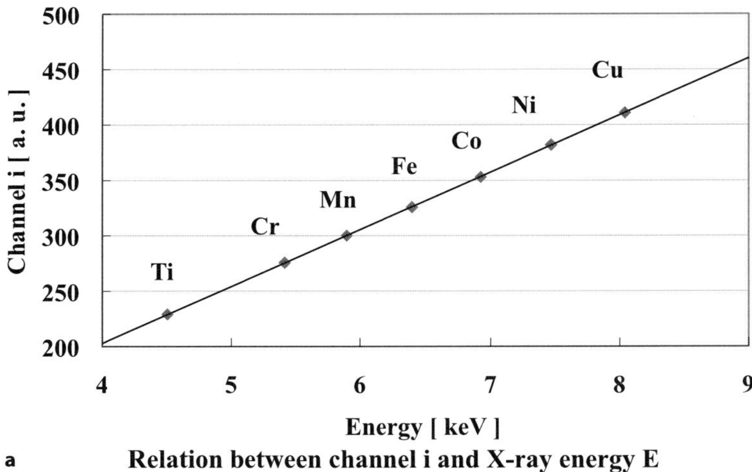

a   Relation between channel i and X-ray energy E

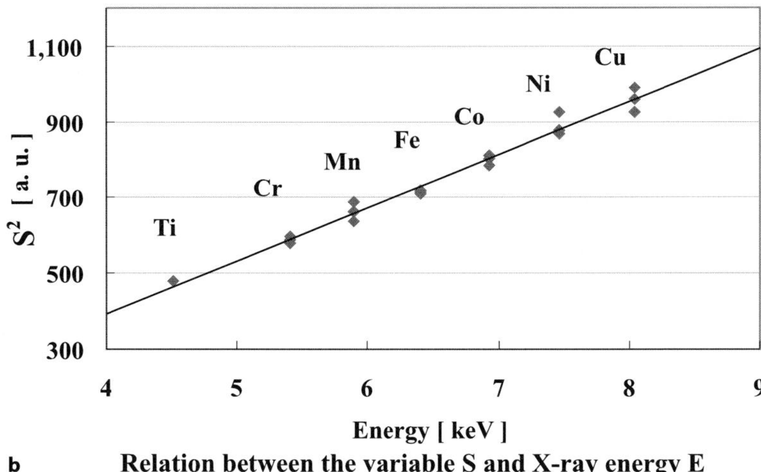

b   Relation between the variable S and X-ray energy E

Fig. 2.12. (a) The relationship between the channel i of the multi-channel analyzer and the X-ray energy E, and (b) the relationship between the variable S that determines the FWHM of the peaks and X-ray energy E. The results were obtained from the analysis using reference samples of Ti, Cr, Mn, Fe, Co, Ni and Cu thin films

shows the relationships between i and E, and S and E, respectively, which were obtained from the analysis using reference samples of Ti, Cr, Mn, Fe, Co, Ni and Cu thin films. Three films whose thicknesses were different (60–280 Å) were analyzed for each kind of the elements. The deviations between measured and calculated values were less than 7% and 1% for Fig. 2.12a,b, respectively.

These expressions are utilized in the multiple peak fitting. Through these relationships, it is possible to determine the position and width of the elemental peaks other than the analyzed reference samples (e.g., P, S, Cl, Ar, K, Ca, Sc, V and Zn).

Multiple Peak Fitting

Based on the position and the width of the peaks obtained in the previous step, the multiple peak fitting is performed to the XRF spectra from the biomedical samples. The peaks are determined by the least squares method and their shapes are represented as Gaussian curves. The single peak is given by

$$y(x) = H \exp \left\{ -\frac{(M' - x)^2}{S'} \right\} \tag{2.25}$$

where M', S' and H determine the position, FWHM and height of the peaks, respectively, and $y(x)$ is the count of the fitted peak in channel x. M' and S' are obtained from the equations (2.22) and (2.23) and are not modulated in this process. The objective function P is represented by

$$P = \sum_{k=1}^{n} \sum_{i} \frac{1}{\omega_i^2} \{C(i) - B(i) - y_k(i)\}^2 \tag{2.26}$$

where $C(x)$ is the experimental counts obtained in channel x, $B(x)$ is the estimated background in channel x and i shows the extent of channels to which the fitting is applied. Y_k is kth peak applied fitting and n is the number of the peaks. The weight function, ω_i, is given by

$$\omega_i = \begin{cases} C(i) - B(i) & (C(i) - B(i) > 10) \\ 10 & (C(i) - B(i) \leq 10) \end{cases} \tag{2.27}$$

This factor enables one to process the large and small peaks at the same time. The minimum value ($= 10$) is set so this factor does not overestimate the small peaks. Then the valuable H is determined for each fitted peak by the steepest decent method. The valuable H is modulated by

$$H_i = H_i - \frac{\partial P}{\partial H_i} \mathrm{d}A, \quad i = 0, 1, \ldots, n \tag{2.28}$$

where H_i is the height of ith peak and n is the number of the fitted peaks. Figure 2.13 shows the example of the multiple peak fitting performed in this procedure and the residual between the measured spectra and the sum of the estimated background and peaks. The black solid and dotted lines show the measured spectra and the estimated background respectively. Blue dotted lines show fitted multiple peaks. The overlapping peaks are decomposed appropriately in accordance with the decrease of the objective function.

After the first fitting, the further fitting can be applied to each peak in the same procedure as the single peak fitting. The positions, width, height of

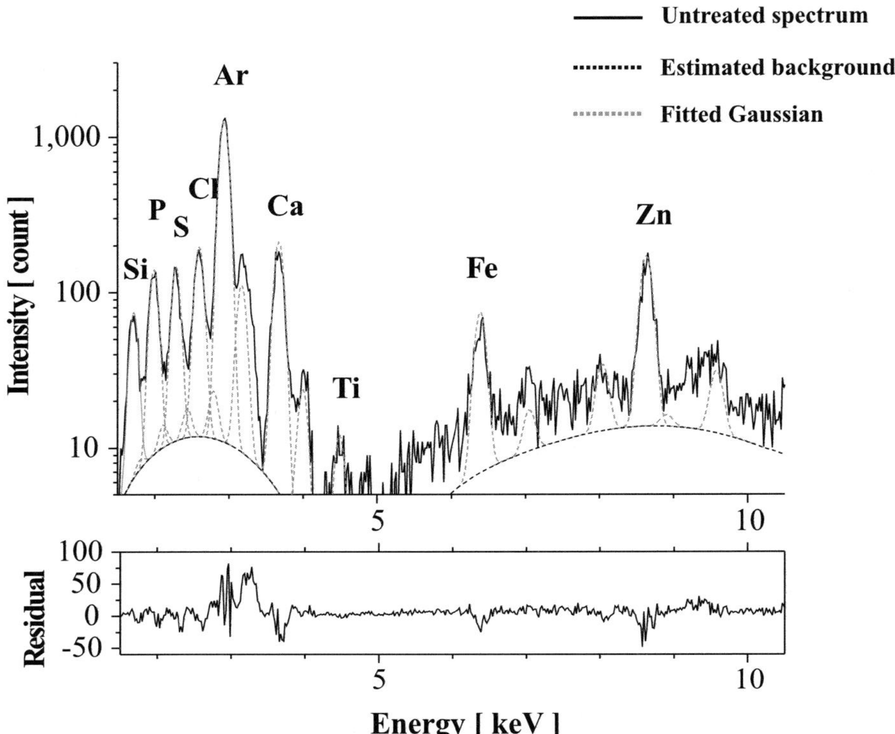

Fig. 2.13. The result and the residual of the multiple fitting applied to the XRF spectra obtained from the biomedical sample. The fitting was performed by the steepest decent method modulating the heights of the peaks

each peak are further adjusted. Figure 2.14 shows the example of the result of further fitting and the residual. The black solid and dotted lines and blue dotted lines show the measured spectra, estimated background and fitted peaks, respectively. It can be seen that the residual has been improved compared to the result from the first fitting. The remained residual is ascribable to the noise and the absorption of fluorescent X-ray in the sample and the path of X-ray.

Once the Gaussians are fitted to all peaks in the XRF spectrum, the net peak areas are obtained for each element. Then the elemental concentrations are calculated according to the equation (2.16).

2.5.2.3 Minimum Detection Limit

If the intensity of fluorescence X-ray is measured under the same conditions, then the matrix effect can be neglected and the relation between the concentration (W_i) and the fluorescent X-ray I_i from element i can be written as follows,

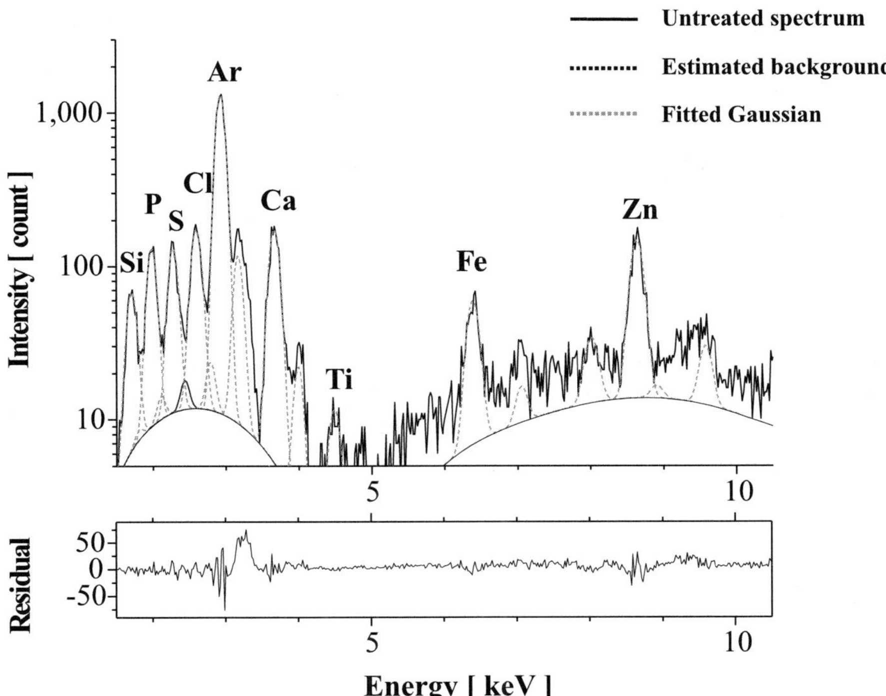

Fig. 2.14. The result and the residual of the further fitting applied after the first multiple peak fitting. The heights, position and width of the peaks were adjusted one by one in the same procedure as the single peak fitting. It can be seen that the residual has been improved compared to the result from the first fitting shown in Fig. 2.13

$$I_i = b_{0i} + b_{1i} \cdot W_i \,. \tag{2.29}$$

Assume now that a reference sample with the element of interest whose concentration is C_S is measured by XRF analysis. If for t seconds the counts of the fluorescent X-ray peak and its background are defined as N_{SP} and N_{SB}, respectively, the slope ($\tan\beta$) of the calibration line is given by

$$\tan\beta = \frac{N_{SP} - N_{SB}}{C_S} \,. \tag{2.30}$$

Using the equation (2.24), the equation (2.23) can be written as

$$N_{SP} = N_{SB} + \tan\beta \cdot C_S \,. \tag{2.31}$$

For an unknown sample, the concentration of the same element W_i can be derived from the peak and background counts N_P and N_B

$$W_i = \frac{N_P - N_B}{\tan\beta} \,. \tag{2.32}$$

The counts in the peak displays Poisson distribution pattern. The standard deviation of the net peak counts "$N_P - N_B$" is given by:

$$\sqrt{N_P + N_B}\,. \tag{2.33}$$

For the peak to be detectable, the following inequality has to be satisfied.

$$\frac{N_P - N_B}{\sqrt{N_P + N_B}} > u_0 \tag{2.34}$$

where $u_0 > 0$. In the region of the peak where the concentration of the element to be detected is at the minimum detection limit, the values of N_P and N_B are almost equal. Thus, the inequality (2.34) can be written as (2.35)

$$N_P - N_B > u_0 \sqrt{2 \cdot N_B}\,. \tag{2.35}$$

From (2.30), (2.32) and (2.35), the minimum detection limit C_{DL} can be expressed as (2.36)

$$C_{\mathrm{DL}} = \frac{u_0 \sqrt{2} \cdot \sqrt{N_B} \cdot C_S}{N_{\mathrm{SP}} - N_{\mathrm{SB}}}\,. \tag{2.36}$$

For 99% confidence limit, the minimum detection limit C_{DL} is defined as (2.37)

$$C_{\mathrm{DL}} = \frac{3 \cdot \sqrt{N_B} \cdot C_S}{N_{\mathrm{SP}} - N_{\mathrm{SB}}}\,. \tag{2.37}$$

It is necessary to display the detection time with C_{DL} because the minimum detection limit changes according to the detection time. In order to make C_{DL} small, the value of the background has to be suppressed and the slope of the calibration curve has to be made larger [16].

2.5.2.4 Conclusion and Discussion

In this section, the computer program developed for the quantitative XRF analysis for the biomedical samples was introduced. Together with the experiments with XRF spectrum analysis, the program is able to fit accurately complex multiplets. This program provides the effective solution to process the data systematically and robustly and has been successfully used both for stringent spectroscopic studies and for routine analysis of large amounts of data. In the following studies, this program was utilized to consider the correlation between the biological functions and the intracellular trace elements. The concentrations of trace elements deduced using this program showed the stability over a wide range of counting statistics.

There are, however, several points to be considered. For the detailed evaluation of the accuracy and reliability of this program, it is necessary to analyze standard materials and to compare the results with the nominal values. In this study the metal thin films made by the vapor deposition on Mylar film, whose thicknesses ranged from 60 to 280 Å, were used as the reference samples. But these thicknesses were measured by the crystal oscillator, and it is possible that they contain certain errors. These samples should be analyzed complementarily by other methods. Furthermore, powerful test of the reliability under the effects of changes in statistics can be made by applying Monte Carlo techniques to simulate spectra of arbitrary intensity [14, 15]. Employing a high-statistics spectrum as the parent probability distribution, one can generate low-statistics spectra.

The absorption of fluorescent X-ray by the sample and the consequent excitation of other elements are not considered in this program. These effects are considered to be small and are neglected because the samples are thin sections and their main constituents are light elements. But in order to reach results with even higher accuracy, it is clear that matrix correction procedures must be adopted individually for the inducing radiation.

2.6 XANES Analysis for Metalloprotein in Biomedical Samples

2.6.1 Principles and Features of Micro-XANES

X-ray absorption fine spectra (XAFS) refer to the oscillatory structure in the X-ray absorption coefficient just above an X-ray absorption edge. It is caused by energy-dependent modulations of photoelectron scattering intensity, which reflects the local atomic structure and chemical information in the analyzed material [17–19]. It provides molecular-level information not previously available using other techniques, such as the species of atoms present and their locations. This information is meaningful in many fields, including material science and biology. The XAFS spectrum is usually divided into the X-ray absorption near-edge structure (XANES) region and the extended X-ray fine structure (EXAFS) region, depending on the strength of photoelectron scattering. The XANES spectrum is represented by the energy region from just below the absorption edge, up to 50 eV above the absorption edge and serves as a site-specific probe of local charge state, coordination, and magnetic moment of the central absorber [20]. This region is determined by full multiple photoelectron scattering, whereas the EXAFS region reflects single and low-order multiple scattering.

XANES spectra are collected in transmission and fluorescence modes. The spectra are represented as the function of photon energy, which is given by

$$\mu = -\exp\frac{I}{I_0} \ (\text{transmission mode}) \ \text{or} \ \frac{I_\mathrm{f}}{I_0} \ (\text{fluorescence mode})$$

where I_0, I and I_f are the incident, transmitted and fluorescent X-ray intensities, respectively. These first two are measured by ionization cambers and the last by Si(Li) detector. A detailed experimental study confirmed that the

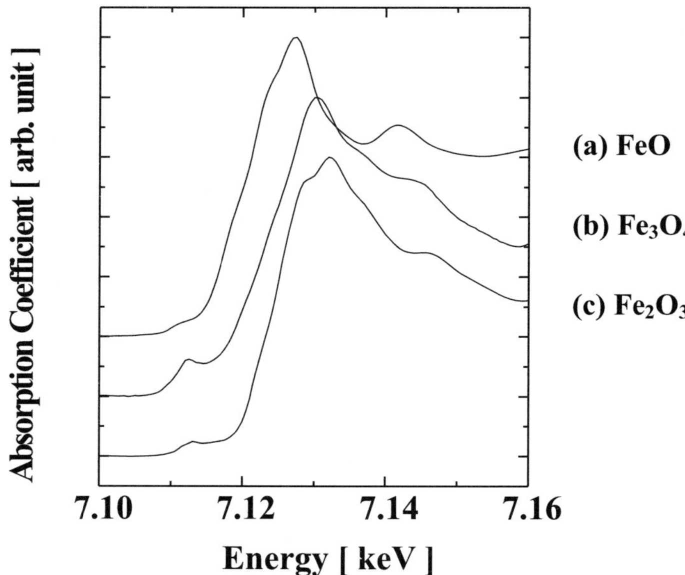

Fig. 2.15. The XANES spectra that were obtained from the powder reference samples of FeO, Fe$_2$O$_3$ and Fe$_3$O$_4$. These spectra were recorded in transmission mode

fluorescent detection mode yielded essentially the same results as the conventional transmission mode for various iron compounds [21]. Fluorescence detection of XAFS is usually the appropriate mode of detection of X-rays for dilute samples, such as biological samples, because the signal to noise ratio in fluorescence mode is often superior to that in transmission mode. Figure 2.15 shows the examples of the XANES spectra that were obtained from powder reference samples of FeO, Fe$_2$O$_3$ and Fe$_3$O$_4$. These spectra were recorded in transmission mode. It can be seen that the absorption edge position shifts to the high-energy side according to the increase of the valence of iron.

The importance of XANES analysis is enhanced in accordance with the increase of new synchrotron X-ray sources. The high spatial resolution, sensitivity and the energy tunability are compatible in the analysis using SR and these characteristics cannot be obtained in other conventional techniques. So micro-XANES analysis is one of the most successful application fields of SR spectroscopy and provides unique information on the local atomic arrangement and the electronic state [22]. The study of the active site of proteins and variations on ligand binding is widely performed in the field of biology, and it is providing very important information for gaining an understanding of the relationship between the protein structure and its function [23]. In this study, the changes in the local chemical structure of iron and zinc binding site, which had been accompanied by the progress of neurodegenerative diseases or the differentiation of mouse ES cells, were investigated by this novel technique.

2.6.2 Beam Line Set-up and Experimental Instruments

In this study, XANES analysis was carried out in SPring-8 BL39XU. The experimental arrangement was the same as that for the XRF analysis, which is shown in Fig. 2.5a. Micro-XAFS measurement is performed using reflection optics in combination with a double crystal monochromator with a constant exit beam. With the double crystal monochromator, the incident energy can be changed continuously. The combination of fundamental/third harmonics of undulator radiation with the Si (1,1,1) reflection of the monochromator enables an energy range from 5 to 37 keV. The incoming photon flux is maximized at every X-ray energy by synchronous tuning between the undulator gap and the monochromator automatically.

References

1. F.R. Elder, A.M. Gurewitsch, R.V. Langmuir, H.C. Pollock, *Phys. Rev.,* **1947**, 71, 829.
2. D.H. Tomboulian, P.L. Harman, *Phys. Rev.,* **1956**, 102, 1423.
3. P. Kirkpatrick, A.V. Baez, *J. Opt. Soc. Am.,* **1948,** 38, 766.
4. C.J. Sparks Jr., *"Synchrotron Radiation Research"*, ed. H. Winick and S. Doniach, **1980**, 459.
5. K.W. Jones, B.M. Gordon, *Anal. Chem.,* **1989**, 61, 341A.
6. K. Janssens, B. Vekemans, F. Adams, P. van Espen, P. Mutsaers, *Nucl. Instrum. Meth. Phys. Res. B,* **1996**, 109, 179.
7. J. Byrne, N. Howarth, *J. Phys. B: At. Mol. Phys.,* **1970**, 3, 280.
8. V. Baryshev, G. Kulipanov, A. Krinsky, *"Handbook of Synchrotron Radiation"* vol. 3, ed. G.S. Brown and D.E. Moncton, **1991**, 639.
9. J. W. Mayer, E. Rimini, *"Ion Beam Handbook for Material Analysis"*, Academic Press, **1977**.
10. J.R. Tesmer, M. Nastasi, J.C. Barbour, C.J. Maggiore, J.W. Mayer, *"Handbook of Modern Ion Beam Materials Analysis"*, Materials Research Society, **1995**.
11. M.A. Kneen, H.J. Annegarn, *Nucl. Instrum. Meth. Phys. Res. B,* **1996**, 109, 209.
12. J.Z. Chu, S.X. Hu, G.Y. Tao, *Chemom. Intell. Lab. Sys.,* **1996**, 32, 83.
13. P.A. Aarnio, H. Lauranto, *Nucl. Instrum. Meth. Phys. Res. A,* **1989**, 276, 608.
14. B. Vekemans, K. Janssens, L. Vincze, F. Adams, P. van Espen, *Spectrochim. Acta B,* **1995**, 50, 149.
15. C.G. Ryan, E. Clayton, W.L. Griffin, S.H. Sie, D.R. Cousens, *Nucl. Instrum. Meth. Phys. Res. B,* **1988**, 34, 396.
16. K. Sugihara, K. Tamura, M. Sato, K. Ohno, *X-ray Spectrometry,* **1999**, 28, 446.
17. P.M. Bertsch, D.B. Hunter, *Chemical Reviews,* **2001**, 101, 1809.
18. C. Bouldin, J. Sims, H. Hung, J.J. Rehr, A.L. Ankudinov, *X-ray Spectrometry,* **2001**, 30, 431.

19. J.J. Rehr, R.C. Albers, *Rev. Mod. Phys.,* **2000**, 72, 621.
20. A.L. Alkudinov, B. Ravel, J.J. Rehr, S.D. Conradson, *Phys. Rev. B,* **1998**, 58, 7565.
21. I. Nakai, C. Numako, S. Hayakawa, A. Tsuchiyama, *J. of Trace and Microprobe Techniques,* **1998**, 16, 87.
22. A. Iida, *X-ray Spectrometry,* **1997**, 26, 359.
23. S. Della Longa, S. Pin, R. Cortes, A.V. Soldatov, B. Alpert, *Biophys. J.,* **1998**, 75, 3154.

3

X-ray Absorption Fine Structure Spectroscopy

3.1 Absorption and Transmission of X-ray through Matter

When an X-ray beam passes through an absorbing medium several important processes take place, namely scattering, ionization, excitation, and the heating and breaking of molecular bonds. The four major processes are photo absorption, coherent scattering (Rayleigh scattering), incoherent scattering (Compton scattering), and pair production. Consequently, the initial beam is absorbed, reflected or transmitted through the matter. The absorption of energy, on the other hand, is directly related to excitation of atoms or molecules of the matter, and emission of characteristic X-rays from the irradiated material.

The attenuation of the X-ray depends on various absorption processes and is well described by a general exponential behavior in nature. Lambert's law states that equal paths in the same absorbing medium attenuate equal fractions of the radiation. Suppose, for the path length dx, the intensity I is reduced by an amount dI. Then, the intensity can be described by equations of the following form:

$$\frac{dI}{I} = -\mu_1 dx, \quad I = I_0 e^{-\mu_1 x}, \tag{3.1}$$

where μ_1 is the linear attenuation coefficient which depends on the state (gas, liquid or solid) of the material. Therefore, it is useful to define the mass attenuation coefficient μ_m, that is independent of the particular phase of the material. Consider an X-ray beam of intensity I with a unit cross-section crossing along a path length of dx in a material with the density ρ. In this layer,

$$\frac{dI}{I} = -\mu_m dm = -\mu_m \rho dx, \tag{3.2}$$

where $m = \rho x$ is the mass per unit area, or area density, in g·cm^{-2}. Equation (3.2) represents the so-called *Bouguer–Lambert–Beer exponential attenuation law.* The mass attenuation coefficient $\mu_m = \mu_l/\rho$ [cm^2/g] is a characteristic value of the substance.

The X-ray attenuation is caused by either absorption or scattering by the atoms of the material. Photoelectric or true absorption (τ) occurs when an inner electron of the atom is completely removed from its shell. The scattering takes place mainly because of the collision of X-ray photons with the loosely bound outer electrons of the atom. Assuming that these two processes are independent, we can express μ_a as the sum of atomic coefficients of photo absorption τ_a and of scattering σ_a,

$$\mu_a = \tau_a + \sigma_a \,. \tag{3.3}$$

Equation (3.3) can be written in greater detail, by taking into account the various attenuation processes, as

$$\sigma_{\text{tot}} = \mu_a = \tau_a + \sigma_{\text{coh}} + \sigma_{\text{incoh}} + \kappa \,, \tag{3.4}$$

where σ_{tot} is the total cross-section per atom in barns/atom (1 barn= 10^{-22} cm^2), τ_a the atomic photoelectric cross-section, σ_{coh} the coherent (Rayleigh) scattering cross-section, σ_{incoh} the incoherent (Compton) scattering cross-section, and κ the pair-production cross section. κ can be neglected in the energy region of synchrotron radiation source.

3.2 X-ray Absorption Fine Structure

When an X-ray beam passes through a medium, a fraction of the X-rays are absorbed in it so that the intensity of incident X-ray beam is attenuated. The relationship between the incident beam intensity I_0 and the transmitted intensity I through an absorbing medium with thickness t is described by Lambert's law,

$$I = I_0 \exp(-\mu\, t) \,, \tag{3.5}$$

where μ is the absorption coefficient. The absorption coefficient for pure element matter generally shows a monotonic decrease with X-ray energy, but with sharp discontinuities characterized by sudden increases at certain energies, corresponding to inner-shell ionization.

When observed with good energy resolution (eV range), these so-called absorption edges exhibit fine structures. In the early days around the 1920s, the structures within 20 eV of the edges were known as a Kossel structure and were attributed to bound excited states. The features observed above the edge beyond about 20 eV were known as the Kronig structure. This division between structures below and above 20 eV from the edge was maintained for a long time. For complex matter, the structure was used to separate the low-energy range strongly affected by the electronic structure of individual atoms from the high-energy range affected by the local spatial structure of neighboring atoms. The Kossel structure of the X-ray absorption fine structure is now called X-ray absorption near-edge structure (XANES), which provides information on the valence state of the elemental atom [1].

The revival of interest in the Kronig structure began with the success of one-electron, single scattering, short-range order theory to explain its oscillatory structure. It was renamed EXAFS (extended X-ray absorption fine structure), and it is now widely used as a tool for the investigation into the local structures in complex materials, though the interpretation of continuum part of the spectra above the photoionization energy in a range of 30 to 60 eV remains unclear (see reviews, for example, Lee et al. [2]). It has been revealed that this part of the X-ray absorption spectrum is originated from the effect of multiple scattering of the excited photoelectron by the local cluster of atomic geometrical arrangements around the absorbing atom. Multiple scattering gives the higher-order atomic correlation functions, while the single scattering gives only the first-order pair correlation function [1].

XANES and EXAFS have now been widely applied to the fields of biology, material science, fundamental physics, and are known generically as X-ray absorption fine structure (XAFS) spectroscopy, carried out by measuring the X-ray absorption of a substance as a function of energy [3].

3.3 EXAFS and XANES

The electromagnetic wave loses its energy because of the interactions with various motions of an atomic nucleus or electrons within a material. X-rays are also absorbed in materials as a result of the interaction with them. If the X-ray beam is irradiated to a certain material, many phenomena can be observed, for example the Compton, elastic and Raman scattering and generation of fluorescence X-rays, photoelectrons and Auger electrons. According to materials and energy of the incident X-ray, the generation probabilities of them have their own values. The X-ray absorption spectra are obtained by measuring the intensities of the incident and transmission X-rays according to the continuous change of the incident X-ray energy. The proportional values of the mass attenuation coefficient, which is $\ln(I_0/I)$, are plotted at each energy of the incident X-ray.

Each element has its own ionization energy. The inner electron jumps out as a photoelectron after the absorption of higher energy X-ray than its ionization energy. Therefore, the absorption of the incident X-ray rapidly increases at the ionization energy. By making the detailed investigation into the absorption spectrum, the fine structure is observed in the energy range from the absorption edge to about 1,000 eV. This fine structure is called XAFS (X-ray absorption fine structure). The fine structure near the immediate vicinity of the X-ray absorption edge is called X-ray absorption near-edge structure (XANES). Beyond the XANES region, starting at about 30–40 eV above the edge absorption maximum, are the fine structure oscillations now generally termed extended X-ray absorption fine structure (EXAFS) [4].

The reason why the fine structure appears in the absorption spectrum is explained by the interference of photoelectrons in the EXAFS region. When an atom absorbs the X-ray energy, the inner shell electrons separate from the binding of the atomic nucleus and jump out as photoelectrons. If the atom is isolated, the photoelectrons are described as the dispersing spherical wave and able to occupy any energy states. Therefore, in this case, the fine structure does not appear in the absorption spectrum. However, in the case of the atom surrounded by other atoms, the probability of the backscattering of photoelectrons is generated and the transition probability is modulated by the interference with it. As a result of it, the fine structure appears in the absorption spectrum. The cyclical structure is not required in the measured sample because EXAFS is the local physics phenomenon around the center atom absorbing X-ray. Therefore, EXAFS analysis can determine the number, kind and distance of the surrounding atoms for the materials that do not display the distinct diffraction patterns, such as amorphous, liquid, molecules and fine grains. The features of EXAFS analysis are described as below:

1. It is able to measure the samples with any phases.
2. It is able to determine the kind and number of surrounding atoms.
3. It is able to determine the distance between the center and surrounding atoms.

On the other hand, EXAFS analysis has some defects:

1. The information obtained by this analysis is limited in the area that is extremely near the center atom (maximum to 5 or 6 angstroms).
2. The procedures of this analysis include voluntariness.
3. It is difficult to measure the light elements.

The origin of XANES is different from that of EXAFS. Because of the electron multiple-scattering process occurring in the XANES range, the spectra do not depend only on atomic distances and coordination numbers, but also, to some extent, on higher-order correlation functions like bond angles or the ratio of distances of the X-ray absorbing atom to its first and second shell neighbor atoms [5]. In the molecular-orbital concept, XANES spectral features are sensitive to the electron structure. On the other hand, in a multiple-scattering one, XANES spectral features are due to the interference effects in the final state by multiple scattering of photoelectrons with neighboring atoms mainly depending on the local geometry. However, the two concepts could not be eventually differentiated because the electronic configuration and the local geometry are closely correlated with each other [6]. The interpretation of XANES is complicated by multiple-scattering effects as well as by the non-validity of the plane-wave-scattering approximation applicable in the EXAFS range.

3.4 Measurement Procedure

In transmission mode, the incident and transmitted X-ray are detected by ionization chambers. They make use of the ionization of the gas within them induced by the incident X-ray beam radiation between the flat plates with several hundreds of voltages. However, the ionization chamber cannot distinguish the energy. Therefore, it is necessary to eliminate the higher harmonics. Beam line 39XU in SPring-8 is equipped with an in-vacuum type undulator and a rotated-inclined, double-crystal monochromator. A combination of the fundamental and third harmonics of the undulator with an Si (111) reflection of the monochromator covers the energy range from 5.7 to 37 keV. A platinum-coated plane mirror is used to reduce the amount of higher harmonics to less than 10^{-4}. The glancing angle can be changed from 2 to 9 μrad with a cut-off energy as low as 8 keV. The signals detected by the ionization chamber are amplified and converted to electric pulses by V-F converter. They are counted and displayed on the screen.

In fluorescence mode, the fluorescence X-ray is detected by SSD. In order to correspond to the absorption coefficient, the intensity of it is divided by that of the incident X-ray beam. The schematic view of beam line and XRF spectrometer at beam line 39XU in SPring-8 are shown in Chap. 2 (Fig. 2.5).

3.5 Experimental Layout for XAFS Analysis

The experimental layout of XAFS analysis is shown in Fig. 3.1. The sample to be analyzed is irradiated with the monochromatic X-ray beam, and the beam energy varied in appropriate small increments to enable observation of the structure. The incident and transmitted beam intensities are monitored by ionization chambers. In the case of low absorption, for example when thin samples are analyzed, $\mu t \cong 0$, the Lambert's equation can be approximated as follows:

$$I_0 - I = I_0 \left(1 - \exp(-\mu t)\right) \cong I_0 \mu t. \tag{3.6}$$

The ejection of the electrons caused by the absorption of X-ray is de-excited by the radiation of fluorescence X-ray and Auger electrons. The emission of the fluorescence X-ray is proportional to the absorption of the incident X-ray. However, the detected fluorescence X-ray is not always proportional to the absorption in case of the thin or condensed sample. If the sample

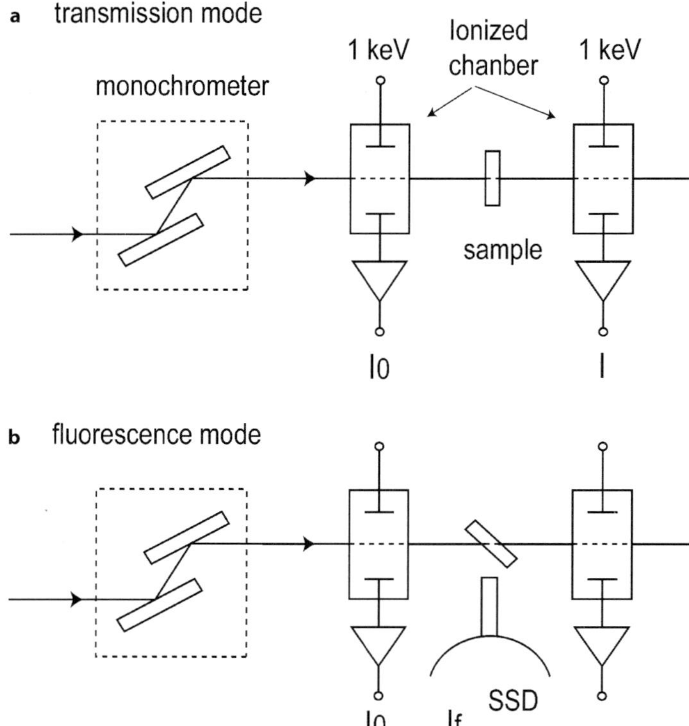

Fig. 3.1a,b. The experimental setup for XAFS. (a) transmission mode, (b) fluorescence mode

is thick enough to neglect its own absorption of fluorescence X-ray and the concentration of the element of interest is low in the sample, the intensity of fluorescence X-ray can be considered to be proportion to the absorption coefficient. This measurement procedure has an advantage to detect trace elements efficiently. By employing the SR source, the detection sensitivity of this method becomes about several thousand times as good as the transmission mode.

The fluorescent intensity I_f emitted from the absorbing medium is proportional to the photon absorption $(I_0 - I)$, the absorption coefficient is given by

$$\mu t = I_f / I_0 . \tag{3.7}$$

The method to measure absorption coefficient using the transmitted photon intensity I is called transmission mode. The alternative mode uses the measurement of the fluorescence intensity I_f of the sample, at the relevant characteristic X-ray peak, and the method is called fluorescence mode.

3.6 Chemical Shift

X-ray absorption near edge structure (XANES) is sensitive to the valence state and neighboring atoms of the absorbing elements, especially the relative position of absorption edge known as chemical shift. K-edge X-ray absorption reflects the transition probability from a $1s$ orbital to the unfilled conduction band. The molecular orbital is sensitive to the binding structure and neighboring atoms of the molecules. Therefore, the density of states of the unfilled conduction band varies with different chemical species, showing as the difference in the X-ray absorption near edge structure. For example, in Fig. 3.2 we show the Fe K-edge XANES spectra of FeO and Fe_2O_3. Using the molecular orbital (MO) theory, one can interpret the peaks in the spectra. For example, peak A is explained as $1s \rightarrow 3d$ transition to $3e_g$ MO, peak B is $1s \rightarrow 4s$ transition to $3a_{1g}$ MO and peak C is $1s \rightarrow 4p$ transition to $4t_{1u}$ MO [7,8]. It is difficult to explain completely the chemical shift that occurred in 3d transition metals, but recent studies using *ab initio* calculation of MO reveal the behavior of electrons in the transitions in X-ray absorption [9].

The effect of oxidation state on chemical shift is significant for biological applications. The chemical state of iron in a biological specimen is complex because the chemical state of iron is a mixture of some kinds of states. However, the chemical shift of X-ray absorption edge structure of iron provides information from which the oxidation state of iron can be obtained. An example of Fe K-edge XANES spectra obtained from a biological specimen is shown in Fig. 3.3. The XANES spectra of reference samples show that the absorption edge of iron with high valence state is shifted to higher energy.

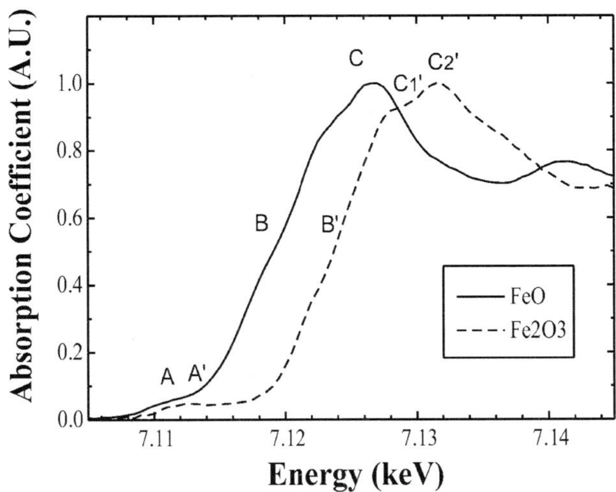

Fig. 3.2. Fe K-edge XANES spectra of FeO and Fe_2O_3. The spectra were normalized by the absorption jump. The measurement was performed in SPring-8 BL39XU

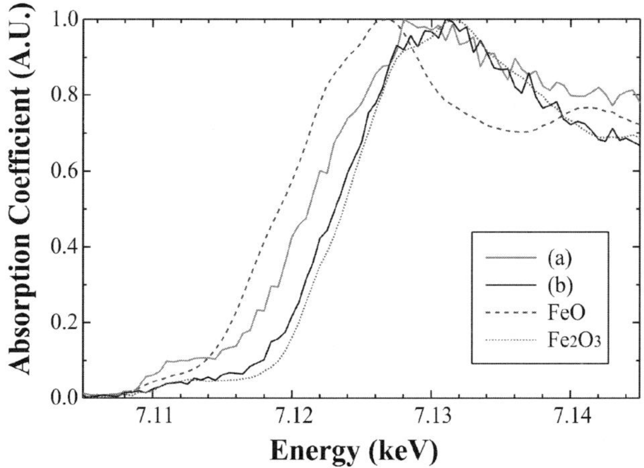

Fig. 3.3. Fe K-edge micro-XANES spectra measured in neuromelanin granules within a surviving neuron (*a*) and in neuromelanin aggregates from a dead neuron (*b*) with Parkinson's disease. This figure is referenced from Yoshida et al. 2001

In Fig. 3.3, the shift of sample (b) is larger than that of sample (a), therefore it can be concluded that iron contained in the sample (b) has higher oxidative state that in sample (a).

3.7 Chemical State Imaging and Selectively Induced X-ray Emission Spectroscopy

X-ray absorption near the edge structure is sensitive to the valence state and neighboring atoms of the absorbing elements as mentioned in Sect. 3.2. The chemical shift of absorption edge can be used for chemical state analysis. If the incident energy near the absorption edge is chosen properly, selective excitation of specific chemical species will occur. Since fluorescent X-rays are emitted accompanied with the excitation of the absorbing elements, the differences of fluorescent yield can be observed in between different chemical species. Here, XANES spectra of FeO (Fe^{2+}) and Fe_2O_3 (Fe^{3+}) are shown in Fig. 3.2. At energies above the absorption edge, such as 7.160 keV, both Fe^{2+} and Fe^{3+} are excited. On the other hand, at energies near the absorption edge, such as 7.120 keV, Fe^{2+} is selectively excited and the excitation of Fe^{3+} is suppressed. Using this phenomenon, one can separate the yields of Fe^{2+} and Fe^{3+} [10]. Using the technique, one can obtain XRF imaging that distinguishes chemical state.

The procedure to separate the fluorescence yields of different chemical states is described as follows. First, we consider the case that iron with the mixed state of Fe^{2+} and Fe^{3+}. In this case, it is necessary to measure the fluorescent yield of the element at two energy point (E_1 and E_2). The following equation is used for calculations to obtain the fluorescence yield separated to Fe^{2+} and Fe^{3+}.

$$\begin{cases} \mu_{\mathrm{Fe2+},E1}\,X_{\mathrm{Fe2+}} + \mu_{\mathrm{Fe3+},E1}\,X_{\mathrm{Fe3+}} = S_{E1} \\ \mu_{\mathrm{Fe2+},E2}\,X_{\mathrm{Fe2+}} + \mu_{\mathrm{Fe3+},E2}\,X_{\mathrm{Fe3+}} = S_{E2} \end{cases} \tag{3.8}$$

where S_{Ej} is the fluorescent yield of the element contained in the sample at energy E_j. $X_{\mathrm{Fe2+}}$ and $X_{\mathrm{Fe3+}}$ are the fluorescent yields contributed by Fe^{2+} and Fe^{3+}, respectively. $\mu_{i,Ej}$ is normalized absorption coefficient at energy $E1$ and $E2$. This is obtained to measure the reference sample with chemical state i at energy Ej or to measure XANES spectrum of the reference sample with chemical state i. We obtain the fluorescence yield separated to Fe^{2+} and Fe^{3+} by solving the equation of the second degree.

When the chemical state of an element contained in a sample is assumed to be superposition of some reference chemical states, we can use the same procedures. Generally, when the number of chemical state is assumed to be n, it is necessary to measure the fluorescent yield at n energy point, as showed below:

$$\begin{pmatrix} \mu_{1,E1} & \mu_{2,E1} & \cdots & \mu_{n,E1} \\ \mu_{1,E2} & \mu_{2,E2} & \cdots & \mu_{n,E2} \\ \vdots & \vdots & \ddots & \vdots \\ \mu_{1,En} & \mu_{2,En} & \cdots & \mu_{n,En} \end{pmatrix} \begin{pmatrix} X_1 \\ X_2 \\ \vdots \\ X_{\mathrm{n}} \end{pmatrix} = \begin{pmatrix} S_{E1} \\ S_{E2} \\ \vdots \\ S_{En} \end{pmatrix} \tag{3.9}$$

where, S_{Ej} is the fluorescent yield of the element contained in the sample at energy E_j, X_i is the fluorescent yield contributed with chemical state i, and $\mu_{i,Ej}$ is the normalized absorption coefficient of chemical state i at energy Ej. This is obtained to measure the reference sample with chemical state i at energy Ej, or to measure XANES spectrum of the reference sample with chemical state i.

We obtain the fluorescence yield separated to different chemical states by solving the equation of the n degree.

References

1. A. Bianconi, *"X-Ray Absorption Principles, Applications, Techniques of EXAFS, SEXAFS and XANES"*, ed. D.C. Koningsberger and R. Prinz, **1988**, 573.
2. P.A. Lee, P.H. Citrin, P. Eisenberger, B.M. Kincaid, *Rev. Mod. Phys.*, **1981**, 53, 769.
3. J.J. Rehr, R.C. Albers, *Rev. Mod. Phys.*, **2000**, 72, 621.
4. G.A. Waychunas, G.E. Brown, M.J. Apted, *Phys. Chem. Minerals*, **1986**, 13, 31.
5. P. Kizler, *Phys. Rev. B,* **1992**, 46, 17.
6. J.H. Choy, D.K. Kim, G. Demazeau, *Phys. Rev. B,* **1994**, 50, 16631.
7. G.L. Glen, C.G. Dodd, *J. Appl. Phys.,* **1968**, 39, 5372.
8. L.A. Grunes, *Phys. Rev. B,* **1983**, 27, 2111.
9. N. Kosugi, *Jpn. J. Appl. Phys.,* **1993**, 32, 13.
10. K. Sakurai, A. Iida, Y. Gohshi, *Adv. In X-ray Anal.,* **1989**, 32, 167.

4

SR Microbeam Analysis at Cellular Level

4.1 Introduction

The advent of microbeam analytical methods has opened up new possibilities of studies of biological systems at the cellular level. In situ techniques enable analysis of living cells in vitro and even in vivo. Electron microprobes and proton microprobes can be used for such studies, but thermal damage to the sample precludes their extensive use. Synchrotron radiation-based microbeams provide virtually an ideal tool for this area of study, offering sensitivity for detection of elements at the parts-per-million level comparable to proton microprobes, but without the adverse effect of sample damage. Although the information is limited to only elemental composition and not the actual compounds, SR also offers the additional advantage of enabling chemical state analysis at trace levels. In this chapter a number of applications of SR microbeam analysis at the cellular level will be presented to illustrate the power of the method.

In the first section, analysis was carried out on cultured single untreated macrophage cells, and those which have been exposed to solutions containing chromium and vanadium. Images of the distribution of a number of elements, and the chemical state of Cr and Fe were obtained in order to gain better understanding of the mechanism and metabolism of phagocytic activities.

In the second section, attempts were made to study single neuron cells.

4.2 Elemental Images of Single Macrophage Cells

4.2.1 Introduction

The interactions and responses of cells (macrophages, neutrophils, fibroblast and endothelial cells) to foreign metal elements have been widely investigated in the past decades. Interaction between cells and metal ions or particles has been investigated (in vivo and in vitro), in the case of wear debris and corrosion in metal implants [1–6]. Phagocytic activity and the inflammatory reaction of macrophages are often discussed from a variety of points of view because of the fact that the macrophage is central to the direction of host inflammatory, immune and phagocytic processes [7]. The problem of cytotoxicity has also been investigated in the interactions between alveolar macrophages and airborne dusts or fly ashes [8–12].

In several previous studies, A.M. Ektessabi et al. [1–3] applied PIXE, microbeam PIXE, and SR-XRF (synchrotron radiation X-ray fluorescence spectroscopy) to the investigation of human tissues around total hip replacements where SUS316L and Ti-6Al-4V had been inserted into the human body for long periods of time. They showed that Fe, Cr, Ni, and Ti were released and distributed into the tissues around the total hip replacement in the form of mechanical friction and corrosion products from the implant. Characteristics of metal particles or ions, such as surface chemistry, surface morphology, net charge, porosity and degradation rate, are critical factors in their interaction with cells. These features can result in the initiation of various responses, such as preferential protein adsorption, complement activation, and cell recruitment [13–15]. However, the mechanisms of phagocytosis of macrophages and cytotoxicity to metal ions or particles are extremely complex, and no single model can fully account for the diverse structures and outcomes associated with particle internalization [16, 17]. While it is possible for each individual response to be identified and analyzed separately, this approach gives an analysis of only a relatively small component of the many factors controlling the overall host responses [18].

In this section we present an investigation by Kitamura and Ektessabi (2000) of fluctuations in the densities and distributions of intracellular elements simultaneously by using SR-XRF imaging technique, on macrophages cultured in a metallic solution environment under differing conditions. XRF analysis is being increasingly utilized in biomedical science [9, 19] because it is, in contrast to most other elemental analysis techniques, capable of analyzing ultra-trace elements nondestructively. In a single cell, metal ions and cellular nutrient density remain more or less constant. This homeostasis is maintained by the delicate balance of transport activities across the plasma and organelle membranes [20].

In this section, interactions between single cells and metal elements are investigated via the measurement of the density fluctuation and also of the localization of those elements, in single cells that have been cultured in a metallic solution environment.

4.2.2 Culture of Macrophages

4.2.2.1 Macrophages

Macrophages are widely distributed throughout the body, displaying great structural and functional heterogeneity. They are to be found in lymphoid organs, the liver, lungs, gastrointestinal tract, central nervous system, serous cavities, bone, synovium, and skin. Macrophage cells play a central role in host inflammatory, immune and phagocytic processes. They are generally large, irregularly shaped cells, measuring 25–50 µm in diameter. They often have an eccentrically placed, round or kidney-shaped nucleus with one or two prominent nucleoli and finely dispersed nuclear chromatin [21].

The aim of the investigation was to clarify the phenomena that occurred within single cells after interactions with foreign metal elements. To achieve that aim, it is necessary to perform the investigations into the intracellular reactions in vivo and in vitro. In particular, the distribution patterns and the densities of the intracellular trace elements are important factors to consider in attempts to understand the functions and defensive mechanisms of the cells against the foreign metal elements. There are only few measurement techniques for investigating the trace elements within single cells. In this study, SR-XRF analysis was employed to measure the trace elements within cultured macrophage single cells. The procedures of the cell culture and the staining, and the morphological observation techniques are described below.

4.2.2.2 Procedures for the Cell Culture

Mouse macrophages (J774.1) were employed in the experiment reported here. These cells were provided by the Riken Cell Bank. The culture medium was RPMI-1640. Fetal bovine serum (FBS) was added into the culture medium. The ratio of FBS and RPMI-1640 was 1 to 10. The cells in DMSO (dimethyl sulfoxide) and culture medium which were frozen by liquid nitrogen, were defrosted in warm water (37 °C). After defrosting, the cells were pipetted into a sterilized tube with the culture medium (5ml). After the centrifugation of the tube, the cells were distributed into a 550 mm diameter dish with 5 ml culture medium. The cells were cultured in an incubation box where the temperature was always kept at 37 °C and the concentration of the carbon dioxide was fixed at 5%. In this case, the cell divisions were observed approximately every eight hours. When the number of cells had increased enough, a fraction was moved into the new dish with a fresh culture medium.

The metallic solutions were prepared as follows. Pure metal compounds of $CrCl_3$, CrO_3, $FeCl_3$ and VCl_3 still enclosed in their bottles were sterilized in an autoclave for 30 min. The metal powders were then dissolved in sterilized PBS solutions. The metallic solutions were sterilized by using a filter with 0.45 µm pores. Metallic solution for the culture is prepared by mixing the culture medium and metallic solution with the ratio of 100 to 1.

4.2.2.3 Histological Observation

Macrophages have ultra-structures within their bodies, such as the nucleus, rough endoplasmic reticulum, Golgi complex and mitochondria. For the inves-

tigation into the intracellular elements, the ultra-structures were made visible
by staining the cells. This is highly desirable since the elemental compositions
may be different among the cell organelles. Giemsa's solution, methyl green
solution, Mayer's hematoxylin solution and eosin Y ethanol solution (0.5%)
were employed in this study for distinguishing the cell nucleus from the cytoplasm.

The staining procedure starts by removing the culture medium from the
dish, followed by rinsing with ethanol (100%), which was poured directly
into the dish. After 10 min, the ethanol was removed and the dish was dried
in air. For Giemsa staining, Giemsa's solution was diluted with the distilled
water (×50) and 5 ml of the diluted solution added into the dish. After half
an hour, Giemsa's solution was removed from the dish. The dish was washed
with water and dried. The light microscopic photograph of macrophage cells
stained by Giemsa's solution is shown in Fig. 4.1. It can be seen that the
nucleoli in the nucleus were stained purple.

For methyl green staining, after rinsing and drying, the dish was wetted
by PBS (phosphate buffered saline) solution and methyl green solution (1ml)
was added. A small dose of acetic acid was pipetted onto the dish. After
10 min, the methyl green solution was removed from the dish followed by

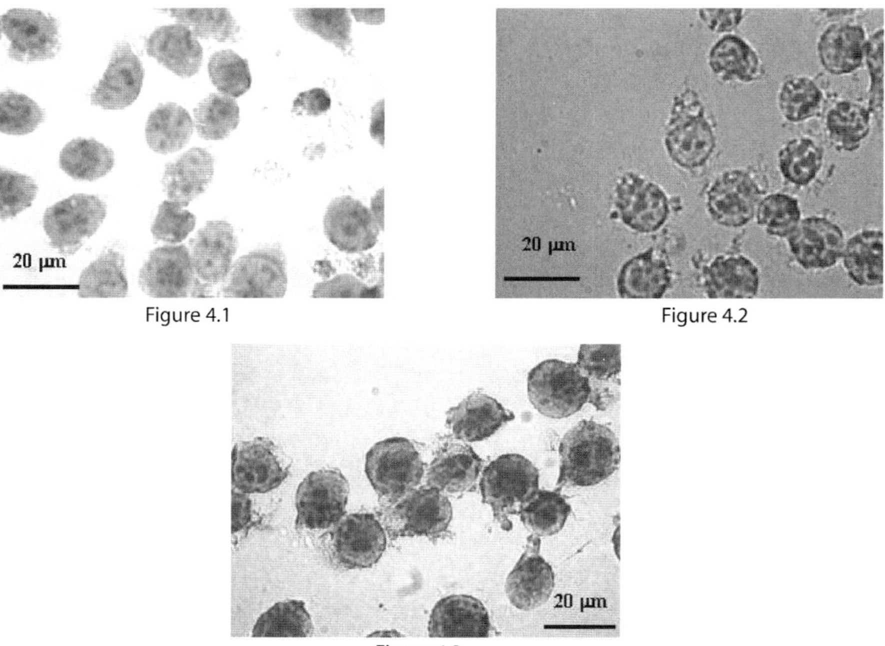

Figure 4.1

Figure 4.2

Figure 4.3

Fig. 4.1, 4.2 and 4.3. Light microscopic photographs of macrophage cells stained
with Gimsa's solution (**4.1**), methyl green solution (**4.2**) and hematoxylin-eosin
solution (**4.3**)

washing in distilled water. The light microscopic photograph of macrophage cells stained by the methyl green solution is shown in Fig. 4.2, where it can be observed that the nucleoli in the nucleus of the macrophage were stained green.

Similar procedure was followed for hematoxylin-eosin staining. After being rinsed and dried, the dish was wetted by phosphate buffered saline (PBS) solution and hematoxylin solution added into the dish. After 10 min, hematoxylin solution was removed from the dish and the dish was washed with water for about 10 min. Eosin Y ethanol solution (0.5%) was diluted with an ethanol solution (80%). The ratio of eosin Y ethanol solution and 80% ethanol solution was 1 to 3. After hematoxylin staining procedures, eosin Y ethanol solution was poured into the dish and a small dose of acetic acid was pipetted onto the dish. After 10 min, the dish was dehydrated by 70% ethanol solution twice and by 98% ethanol solution twice also. The light microscopic photograph of macrophage cells stained by hematoxylin-eosin solutions is shown in Fig. 4.3. It can be observed that the nucleoli in the nucleus of the macrophage were stained blue and other parts were stained red.

By performing the staining procedures, many of the cell organelles can be visualized and recognized. The staining solutions do not contribute any contaminants that can interfere with the measured elements, and thus SR-XRF analysis can be carried out precisely on selected cell organelles.

The staining procedures described above were employed in order to make the cell organelles visible. Another staining procedure, the trypan blue, is used to distinguish the dead cells from the living ones. If the cells are alive, they are not stained by trypan blue solution, but if they are dead the solution enters the cells and the cells are stained blue.

For this investigation, the cells were cultured in solutions containing different concentrations of chromium chloride ($CrCl_3$), chromium oxide (CrO_3), iron chloride ($FeCl_3$) and vanadium chloride (VCl_3). The dosing solutions were prepared by dissolving $CrCl_3$ (0.04 and 0.4 g/L), CrO_3 (0.04 and 0.4 g/L), $FeCl_3$ (0.04 and 0.4 g/L) and VCl_3 (0.04 and 0.4 g/L) in the culture medium. After exposure times of 0.17, 0.5, 4, 8, 12, 24, 48 and 72 h, the cells were stained by trypan blue solution. In order to evaluate the strength of the cytotoxicity of the metallic elements, it is necessary to define a parameter: "death rate", defined as the ratio of the number of the stained cells divided by the total cells sampled. If the membrane and intracellular structures of the cell are broken and it is impossible to judge the color, the death rate is regarded as 100%.

The resulted for culture in Cr chloride ($CrCl_3$) solutions (0.04 and 0.4 g/L) under different conditions are shown in Table 4.1 and 4.2. In these cases, the death rate of the cells was about zero under all conditions. It is probable that the macrophage cells might be all alive in a Cr solution under these concentrations. However, it cannot be concluded that all of the macrophage cells survived in these Cr solutions due to the fact that the trypan blue dye exclusion test cannot judge the death of the cells perfectly.

Table 4.1. Rate of death in macrophage cells cultured in $CrCl_3$ solution (0.04 g/L)

Time (h)	Num. of dead cells	Num. of sample cells	Death rate (%)
0.17	1	295	0.34
0.5	0	200	0.00
4	0	200	0.00
8	0	200	0.00
12	0	200	0.00
24	0	200	0.00
48	0	200	0.00
72	0	200	0.00

Table 4.2. Rate of death in macrophage cells cultured in $CrCl_3$ solution (0.4 g/L)

Time (h)	Num. of dead cells	Num. of sample cells	Death rate (%)
0.17	0	200	0.00
0.5	0	200	0.00
4	0	200	0.00
8	0	200	0.00
12	0	200	0.00
24	0	200	0.00
48	0	200	0.00
72	0	200	0.00

Table 4.3. Rate of death in macrophage cells cultured in CrO_3 solution (0.04 g/L)

Time (h)	Num. of dead cells	Num. of sample cells	Death rate (%)
0.17	3	303	0.99
0.5	0	200	0.00
4	0	200	0.00
8	54	239	22.59
12	67	220	30.45
24	147	227	64.76
48	358	423	84.63
72	200	200	100.00

Tables 4.3 and 4.4 show the results for culture in Cr oxide (CrO_3) solutions (0.04 and 0.4 g/L) under different conditions. In Table 4.3, after 8 h exposure, the number of the cells stained blue increased. The death rate of the cells reached 100% after 72 h. In Table 4.4, the cells started to be stained after 4 h. The death rate of the cells reached 100% after 24 h.

Table 4.4. Rate of death in macrophage cells cultured in CrO_3 solution (0.4 g/L)

Time (h)	Num. of dead cells	Num. of sample cells	Death rate (%)
0.17	0	200	0.00
0.5	0	200	0.00
4	28	310	9.03
8	49	270	18.15
12	29	195	14.87
24	200	200	100.00
48	200	200	100.00
72	200	200	100.00

Table 4.5. Rate of death in macrophage cells cultured in VCl_3 solution (0.04 g/L)

Time (h)	Num. of dead cells	Num. of sample cells	Death rate (%)
0.17	0	200	0.00
0.5	0	200	0.00
4	0	200	0.00
8	0	200	0.00
12	0	200	0.00
24	0	200	0.00
48	0	200	0.00
72	0	200	0.00

Table 4.6. Rate of death in macrophage cells cultured in VCl_3 solution (0.4 g/L)

Time (h)	Num. of dead cells	Num. of sample cells	Death rate (%)
0.17	0	200	0.00
0.5	0	200	0.00
4	0	200	0.00
8	0	200	0.00
12	0	200	0.00
24	69	200	26.45
48	200	200	100.00
72	200	200	100.00

Therefore, it can be concluded that the cells died in a shorter time and
the death rate of the cells reached 100% rapidly when the concentration
of the Cr oxide solution is higher, implying that the concentration of Cr
oxide solution is related to the strength of the toxicity to the cells. Furthermore, the rapid increase of the death rate in high concentration Cr

Table 4.7. Rate of death in macrophage cells cultured in $FeCl_3$ solution (0.04 g/L)

Time (h)	Num. of dead cells	Num. of sample cells	Death rate (%)
0.17	0	200	0.00
0.5	0	200	0.00
4	0	200	0.00
8	0	200	0.00
12	0	200	0.00
24	0	309	0.00
48	0	200	0.00
72	0	200	0.00

Table 4.8. Rate of death in macrophage cells cultured in $FeCl_3$ solution (0.4 g/L)

Time (h)	Num. of dead cells	Num. of sample cells	Death rate (%)
0.17	2	200	1.00
0.5	3	200	1.50
4	0	200	0.00
8	13	192	6.77
12	25	200	12.50
24	0	200	0.00
48	200	200	100.00
72	200	200	100.00

solution may be explained by the phenomenon that the cell divisions can be interrupted by excessive uptake of Cr. In these cases, the cells were severely affected by the cytoxicity of Cr oxide. Compared with the results of exposure to Cr chloride, it can be concluded that the valence state ($+3$ or $+6$) of Cr in the solution is related to the toxicity to the macrophage cells.

Tables 4.5 and 4.6 show the results for culture in V chloride (VCl_3) solutions (0.04 and 0.4 g/L) under different conditions. All of the cells were alive when the dosing solution concentration was 0.04 g/L. For the 0.4 g/L concentration solution, the death rate of the cells started to increase after 12 h. Under this high dose of V solution, the cells were injured by the toxicity of V.

For the exposure to Fe chloride ($FeCl_3$) solutions (0.04 and 0.4 g/L) the results are shown in Tables 4.7 and 4.8. All cells were alive when the concentration of the dosing solution was 0.04 g/L.

The graphs of the death rate according to the exposure time are shown in Figs. 4.4 and 4.5. Based on the trypan blue dye exclusion test, the cytotoxicity of metallic elements varies for different metals. Furthermore, there is evidence

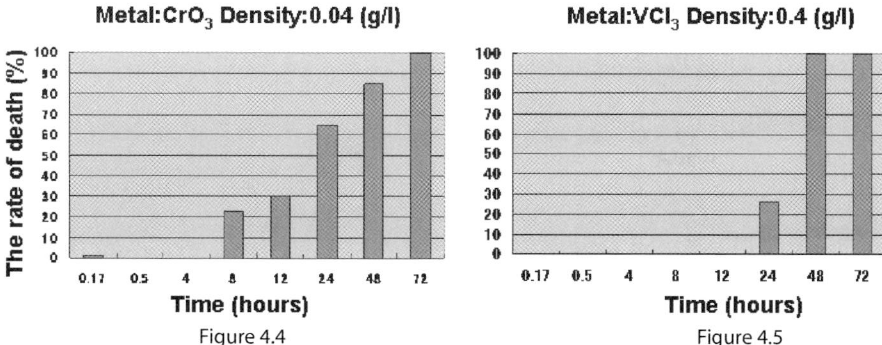

Fig. 4.4 and 4.5. Graphic of the death rate vs. time in macrophage cells cultured in 0.04 g/L CrO₃ (**4.4**) and 0.4 g/L VCl₃ (**4.5**)

that the valence state of the metallic element affects cytotoxicity. However, considering that the trypan blue dye exclusion test does not indicate dead or alive cells perfectly, it is probable that the cytotoxicity of metallic elements may be more complex. This necessitates application of other analytical techniques for further investigations into the intracellular phenomena at a single cell level.

4.2.2.4 Morphological Observation

When macrophage cells internalize the foreign bodies, many ruffles are observed on the surface of the cells. It is possible that the surface changes with the process of cell death due to excessive accumulation of metallic element within their bodies. In order to observe the differences of the morphological states of the single cells after the culture in the normal and metallic solution environment, scanning electron microscope (SEM) analysis was employed.

Silicon plates were employed as substrate because of their good electrical conductivity. The plates were immersed in ethanol solution and sterilized by ultraviolet ray for 24 h. Macrophage cells were cultured on these silicon plates and fixed by a formalin solution for 24 h. The fixed cells on the silicon plate were then coated with a thin layer of carbon. The selection of the fixation solution is important to visualize the fine structures on the surface of the cells clearly. For comparison, other fixing techniques were tried, namely by ethanol and methanol. The results are shown in Figs. 4.6, 4.7 and 4.8. When the cells were fixed by methanol or ethanol solutions, the fine structures on the surface of the cell were lost. However, with formalin the fine structure of the cell was distinct as shown in Fig. 4.8.

The cells cultured under normal condition are shown in Figs. 4.9 and 4.10. The cells cultured in a Cr chloride solution (0.4 g/L) for 1.5 h are shown in Figs. 4.11 and 4.12. The cells cultured in V chloride solution (0.04 g/L) for 1.5 h are shown in Figs. 4.13 and 4.14. The cells cultured in Cr oxide so-

lution (0.04 g/L) for 4 h are shown in Figs. 4.15 and 4.16. From the results
of trypan blue dye exclusion test, it can be concluded that the toxicity of
a Cr oxide solution is the strongest and that of the Cr chloride solution
is the weakest of all solutions employed in that experiment. It can be seen
that the structure on the surfaces of the cells cultured in a Cr oxide solution (0.04 g/L) for 4 h were different from the cells cultured in the other
metallic solutions. The cells shown in Figs. 4.15 and 4.16 may be injured
within their bodies because of the unusual activities originated from the uptake of the excessive toxic metal. When some foreign bodies are internalized
in macrophage cells, the ruffle structures are observed on the surface of the
cells. However, in this case (Figs. 4.15 and 4.16), the change of the shape
of the cell may be different from that of the cell internalizing some foreign

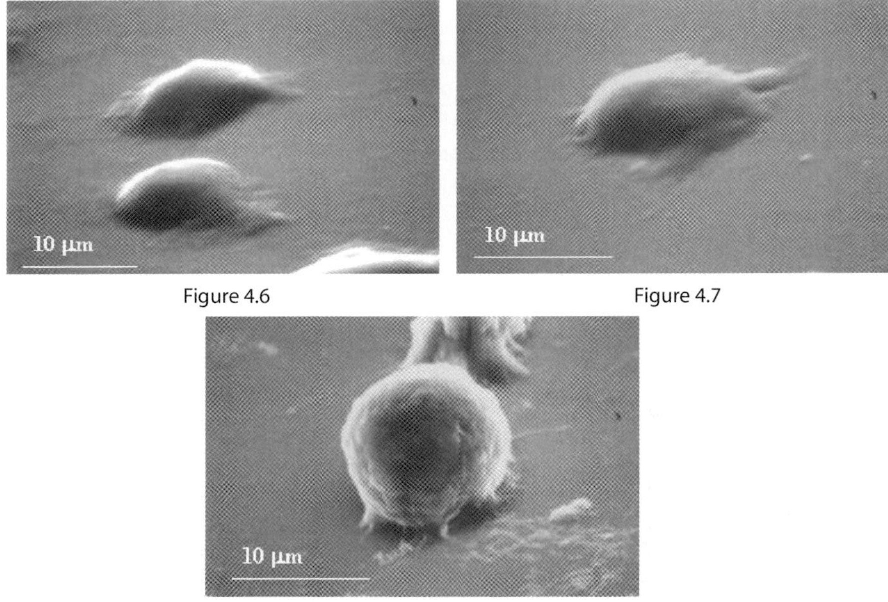

Figure 4.6

Figure 4.7

Figure 4.8

Fig. 4.6, 4.7 and 4.8. In order to compare formalin fixation technique with other
fixation solution, macrophages were fixed with ethanol (**4.6**), methanol (**4.7**) and
formalin (**4.8**). It can be seen that the fine structure of the cells was lost when the
cells were fixed with ethanol or methanol. However, it is clear that the fine structure
of the surface was distinct in (**4.8**)

Fig. 4.9–4.16. Figure (**4.9**) and (**4.10**) show cells cultured under normal conditions. The cells cultured in 0.4 g/L CrO_3 solution for 1.5 h are shown in (**4.11**) and
(**4.12**). The cells cultured in 0.04 g/L VCl_3 solution for 1.5 h are shown in (**4.13**)
and (**4.14**). And the cells cultured in 0.04 g/L CrO_3 solution for 4 h are shown in
(**4.15**) and (**4.16**)

bodies. The states of the surface of the cells were almost identical when the cells were cultured in Cr chloride, V chloride for 1.5 h and non-metallic solutions.

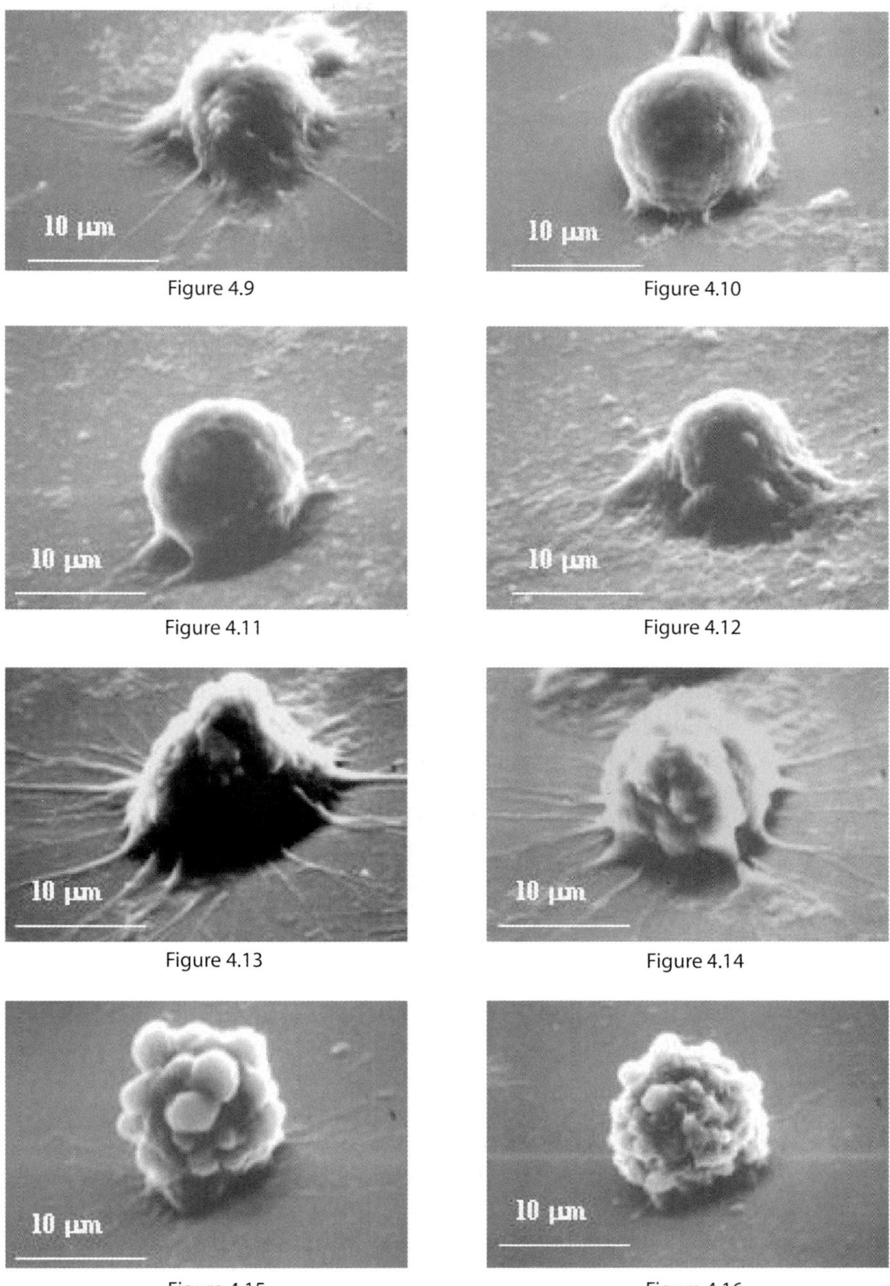

Figure 4.9

Figure 4.10

Figure 4.11

Figure 4.12

Figure 4.13

Figure 4.14

Figure 4.15

Figure 4.16

It can be concluded that the structure of the surface of macrophages approximately reflects the intracellular activities against foreign metal elements. Morphological observation, thus, will assist the investigations into cell functions and intracellular activities.

4.2.3 SR Measurement

4.2.3.1 Sample Preparation

For *treated* macrophage samples, the cells were exposed to solutions containing different concentrations of Cr chloride, Cr oxide, V chloride and Fe chloride during culture. For *untreated* macrophage samples, the cells were cultured under normal conditions without metal uptake. The J774.1 mouse macrophages were provided by the Riken Cell Bank. The culture was carried out on 550 mm dishes in RPMI-1640 medium supplemented with 10% fetal bovine serum. After exposure to metallic solutions, the cells were washed with the culture medium and separated by centrifuge for 5 min. After centrifugation, the cells were immersed in 70% ethanol for 4 h. These procedures were repeated three times and then the cells were immersed in 100% ethanol for 1 h. The ethanol solution containing the cells was pipetted on PET films and dried for XRF measurement.

4.2.3.2 Experimental Set-up

The X-ray fluorescence spectra and elemental images were mainly obtained at beam line 4A in the High Energy Accelerator Institute in Tsukuba, using the X-ray beam emitted by the 2.5 GeV storage ring "Photon Factory". This beam line is a hard X-ray beam line with bending magnet and is dedicated to X-ray analysis in material science and biology, and ultra-trace elemental analysis. A multilayer monochromator provides X-rays with energy that can be varied from 5.0 to 30 keV. The X-ray beam obtained in the experimental hutch has energy resolution $\Delta E/E$ of less than 1×10^{-4}, photon flux of $\sim 10^{11}$ photons/s, and beam size of $1(V) \times 3(H) \, mm^2$. A platinum-coated plane mirror is used to reduce higher harmonics to less than 10^{-4}. A slits system is used to limit the vertical and horizontal beam width incident beam to improve resolution, and a KB (Kirkpartrick–Baez Optics) focusing mirror is located in front of the sample. The final beam size was 6 μm in diameter in this study. Fluorescent X-ray was detected by SSD (Si(Li)) in air.

Some of the X-ray fluorescence and most of the absorption spectra were measured at the Japan Synchrotron Radiation Research Institute using the X-ray beam emitted by an 8 GeV storage ring "SPring-8". Beam Line-39XU in SPring-8 was used to do XRF and XANES analysis. This beam line is a hard X-ray undulator beam line and is mainly used for studying X-ray absorption, X-ray microanalysis in material and biology, and ultra-trace elemental analysis. It consists of an in-vacuum undulator with 32 mm period length and 140 period number, and it generates a horizontally polarized X-rays beam. Combination of the undulator harmonics and a Si (111) double crystal monochromator provides X-rays with an energy range from 5.7 to 37 keV. The X-ray beam obtained in the experimental hutch has the energy resolution $\Delta E/E$ of less than 2×10^{-4}, photon flux of $\sim 10^{12}$ photons/s, and beam size of $0.5(V) \times 1.3(H)\,\text{mm}^2$. A platinum-coated plane mirror is used to reduce higher harmonics to less than 10^{-4}. An incident beam slit is used to limit vertical and horizontal beam width to improve resolution, and a pinhole $10\,\mu\text{m}$ in diameter is located between the incident beam slit and the sample. Fluorescent X-ray was detected by SSD (Si(Li)) in vacuum.

The detailed set-up of the beam lines is described previously and are shown in Fig. 2.5.

4.2.4 Experimental Results

4.2.4.1 Elemental Images of Macrophages

Elemental Distribution in Untreated Macrophages

A microscopic photograph of a macrophage dried on PET film is shown in Fig. 4.17a. An XRF spectrum measured at a point in the cell with a high density of Ca, Zn, K and P is shown in Fig. 4.18. The main intracellular elements detected were P, S, Cl, K, Ca and Zn. Ca has the highest intensity in the X-ray energy range of 1 keV to 10 keV.

The elemental distributions of Ca and Zn within a single *untreated* macrophage are shown in Fig. 4.17b and 4.17c, respectively. While the distribution of Ca and Zn in the macrophage cell are almost identical, the densities are completely different.

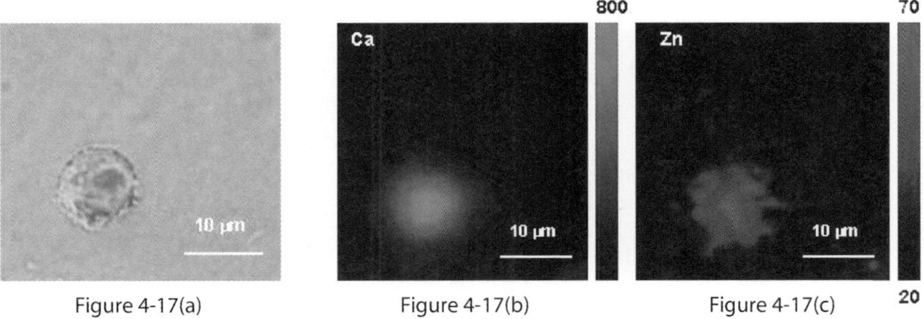

Fig. 4.17. A microscopic photograph of a macrophage dried on PET film is shown in (**a**). The elemental distribution of Ca and Zn within a single *untreated* macrophage are shown in (**b**) and (**c**), respectively. These images are matrices of 30×30 pixels of $1\,\mu\text{m}$ resolution. The measurement time was $5\,\text{s}$

Elemental Distribution in Treated Macrophages,
Exposure to Cr Chloride Solution

Treated macrophages were cultured in a Cr chloride solution environment
with different exposure times to investigate the differences in uptake of Cr
and also the variations in the distribution of the internalized elements and
intracellular elements. The exposure times were 0.17, 0.5, 4, 12, 24 and 48 h.
Seven images for each sample of cells cultured in the Cr chloride solution
environment showing the distributions of P, S, Cl, K, Ca, Cr and Zn were
obtained with the exposure time as the variable. The images (elemental distributions) of Ca, Zn and Cr for four typical cases are shown in Fig. 4.19a–d.
The distributions of Ca, Zn and Cr that are shown in Fig. 4.19a, b, c and d
were measured in cells that were cultured in a Cr solution of 0.04 g/L medium
for 0.17, 0.5, 4 and 24 h, respectively. These images are 1 μm resolution. The
density of Ca, Cr and Zn are displayed as shades of green, red and blue,
respectively, as shown in Fig. 4.19a–d. In Fig. 4.19a_5 and b_5, the images
of Ca and Cr are overlapped. The overlapping areas are expressed by their
superimposed colors.

In the image shown in Fig. 4.19a, the ranges of the fluorescent X-ray intensities of Cr, Ca, and Zn are from 0 to 45, 0 to 107 and 0 to 26 counts, respectively. The measurement time was 5 s. The distribution patterns shown in
Fig. 4.19a are typical images that can be seen in the initial stage of the uptake
of the foreign metal element. The experimental results shown in Fig. 4.19b
and 4.19c reveal that Ca and Zn have almost identical distribution pattern
to that of *untreated* cells, as can be seen by comparing with Fig. 4.19a.

In the image shown in Fig. 4.19b, the ranges of the fluorescent X-ray intensities of Cr, Ca and Zn are from 0 to 126, 0 to 966 and 1 to 155 counts,
respectively. The measurement time was 5 s. The distribution patterns shown
in Fig. 4.19b are typical images observed when the uptake of Cr into the cell
is increasing and the distributed Cr populations within the cell are concentrated. It is assumed that this distribution pattern observed is independent
of the exposure time. Ca and Zn are not always correlated. As shown in

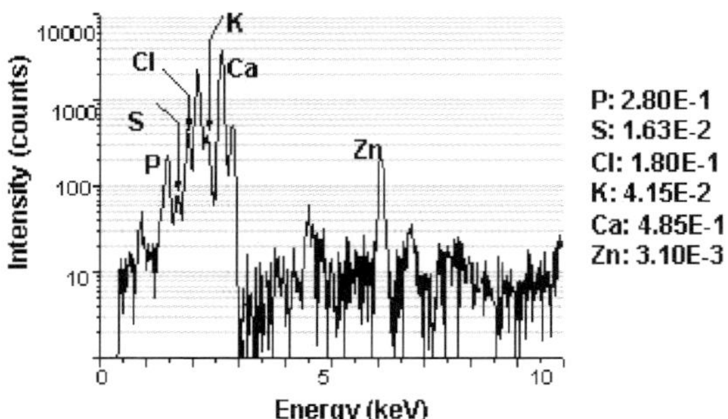

Fig. 4.18. An XRF spectrum of a single *untreated* macrophage measured at the
point with the highest intensity of Ca, Zn, K and P

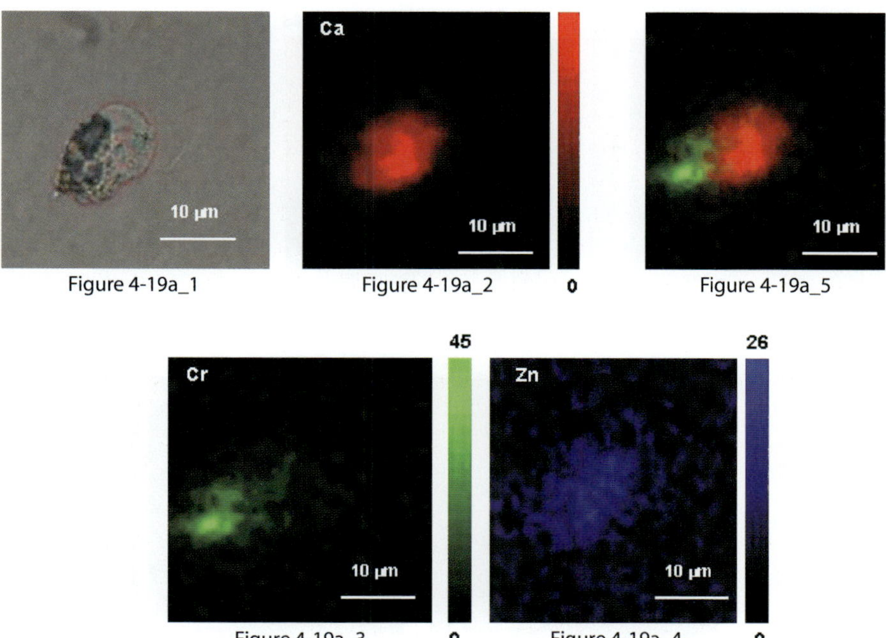

Fig. 4.19a_1–a_5. A microscopic photograph of a dried macrophage dried on PET film is shown in (**a_1**). The elemental distribution of Ca, Cr and Zn within a single *treated* macrophage cultured in 0.04 g/L Cr solution for 0.17 h are shown in (**a_2**), (**a_3**) and (**a_4**), respectively. These images are matrices of 42×42 pixels of 1 µm resolution. In (**a_5**), the images of Ca and Cr are overlapped

Fig. 4.20, Zn and Ca are completely separated from each other and Zn is localized near Cr in the cell.

In the image shown in Fig. 4.19c, the ranges of the fluorescent X-ray intensities of Cr, Ca and Zn are from 0 to 11, 0 to 21 and 0 to 18 counts respectively, obtained in 5 s. When the uptake of Cr is at the low level, Cr, Ca and Zn have almost identical distribution patterns. However, the densities of Ca and Zn within the cell are much lower than those within the control (*untreated*) cells. The distributions of Ca and Zn become almost identical patterns independent of the exposure time for low level Cr uptake.

In the image shown in Fig. 4.19d, the ranges of the fluorescent X-ray intensities of Cr, Ca and Zn are from 0 to 7000, 0 to 280 and 0 to 180 counts respectively, obtained in 5 s. The distribution patterns of Cr, Ca and Zn become identical with the high densities of these elements, when the uptake and accumulation of Cr are high. It is likely that the uptake of Cr was saturated and the cell division and other activities ceased.

When Cr chloride solution was exposed to macrophages, four typical distribution patterns of Cr, Ca and Zn were observed. In Fig. 4.19b, the densities

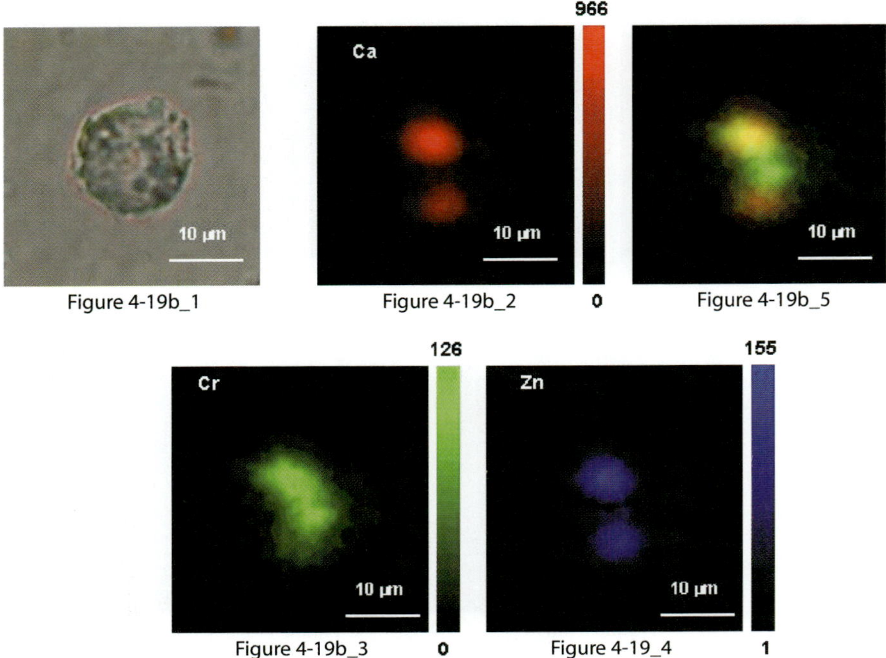

Figure 4-19b_1   Figure 4-19b_2   Figure 4-19b_5

Figure 4-19b_3   Figure 4-19_4

Fig. 4.19b_1–b_5. A microscopic photograph of a dried macrophage dried on PET film is shown in (**b_1**). The elemental distribution of Ca, Cr and Zn within a single *treated* macrophage cultured in 0.04 g/L Cr solution for 0.5 h are shown in (**b_2**), (**b_3**) and (**b_4**), respectively. These images are matrices of 42×42 pixels of 1 μm resolution. In (**b_5**), the images of Ca and Cr are overlapped

of the intracellular Ca and Zn increased and their distribution patterns were localized within the cell. This pattern was unusual. A more typical pattern is shown in Fig. 4.19c, where Cr, Ca and Zn were not localized but distributed and their densities were low within the cell. This pattern was also observed for samples exposed to a Cr chloride solution (0.04 g/L) from 0.17 to 24 h. The meaning of this state on the process of phagocytosing foreign metal elements is considered at the end of this section.

Quantification of Density of Intracellular Elements as a Function of Time and Dose

For the quantification of the densities of the intracellular elements, XRF spectra were obtained at a point in each cell with the highest densities of the elements. The measurement time was 200 s. for each point. In order to explain the increase or decrease in the density of the elements in relation to the uptake of Cr, it is useful to define a parameter, namely "relative density" for each element expressing the ratio of the intensities of that element in the *treated* macrophage divided by that in the *untreated* macrophage. Therefore,

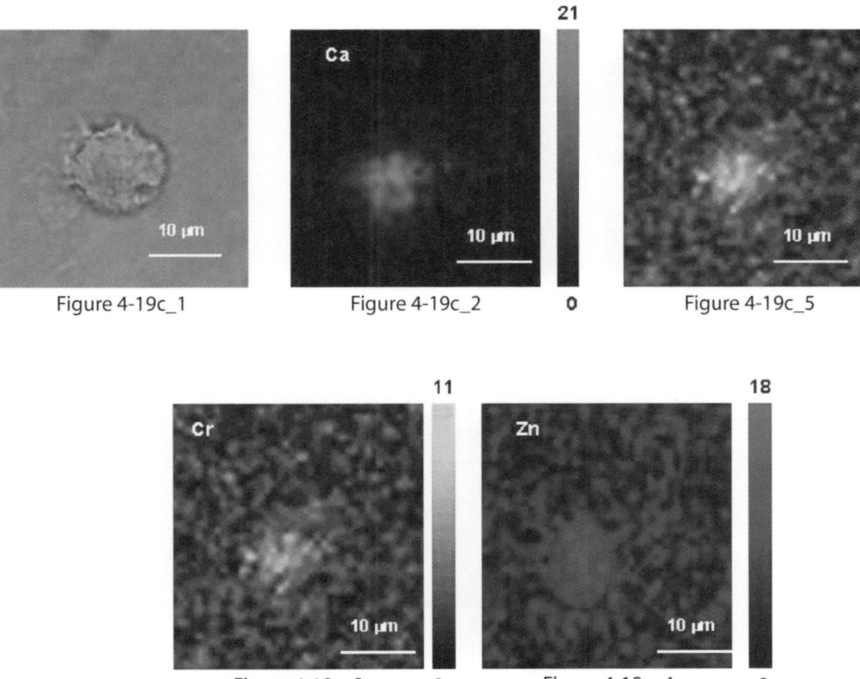

Fig. 4.19c_1–c_5. A microscopic photograph of a dried macrophage dried on PET film is shown in (**c_1**). The elemental distribution of Ca, Cr and Zn within a single *treated* macrophage cultured in 0.04 g/L Cr solution for 4 h are shown in (**c_2**), (**c_3**) and (**c_4**), respectively. These images are matrices of 42×42 pixels of 1 μm resolution. In (**c_5**), the images of Ca and Cr are overlapped

a relative density of less than 1 in a cell means that the density of that element in the *treated* cell decreases due to the uptake of Cr, and conversely a value greater than 1 corresponds to an increase.

Effect of Cr Uptake on P Density

The spectrum shown in Fig. 4.21 was obtained from the cell shown in Fig. 4.19a with the highest density of Ca, cultured in a Cr chloride solution (0.04 g/L) for 0.17 h. The value of the relative density for P was 0.636. Values for other cells exposed to the same solution for the same time of 0.17 h are almost identical. Although the cell was exposed to Cr chloride solution for only 0.17 h, the P may be emitted from inside the cell and bonding to some other molecules.

The spectra in Fig. 4.22 were obtained from the cell shown in Fig. 4.19b, which has the highest densities of Cr and Ca and was cultured in a Cr chloride solution (0.04 g/L) for 0.5 h. The resulting relative density of P was 2.64. The amount of the internalized Cr was large in spite of the short exposure

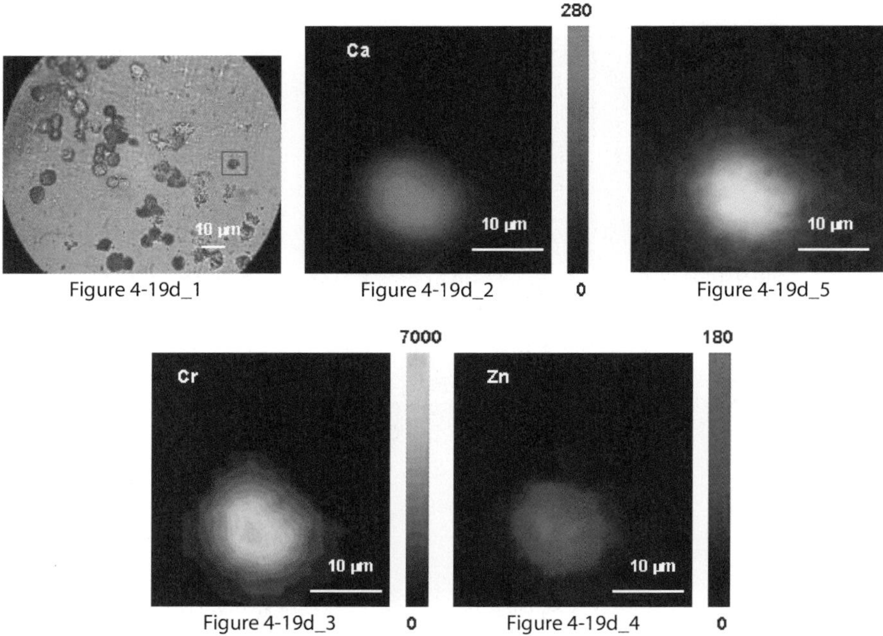

Fig. 4.19d_1–d_5. A microscopic photograph of a dried macrophage dried on PET film is shown in (**d_1**). The elemental distribution of Ca, Cr and Zn within a single *treated* macrophage cultured in 0.04 g/L Cr solution for 24 h are shown in (**d_2**), (**d_3**) and (**d_4**), respectively. These images are matrices of 31 × 31 pixels of 1 µm resolution. In (**d_5**), the images of Ca and Cr are overlapped

time. It appears that the rapid inflow of Cr into the cell induced subsequent inflow of P.

The spectra in Fig. 4.23 were obtained from the cell shown in Fig. 4.19c with the highest densities of Cr and Ca, cultured in a Cr chloride solution (0.04 g/L) for 4 h. The value of the relative density of P was 0.134. It appears that when the amount of the internalized Cr into the cell was small, the density of P was also small, and unrelated to the exposure time.

The spectra in Fig. 4.24 were obtained from the cell shown in Fig. 4.19d with the highest densities of Cr and Ca, cultured in a Cr chloride solution (0.04 g/L) for 24 h. The value of the relative density of P was 1.74.

The following mechanism appears plausible. First, the density of P within the cell decreased after exposing the cell to a Cr chloride solution. If the up-taken Cr into the cell is low, the density of P within the cell may be maintained at low level independent of the exposure time. However, the intracellular density of P increases when the density of Cr within the cell is large regardless of the rate of inflow of Cr into the cell. It appears that the density fluctuation of P is correlated with the uptake of Cr into the cell.

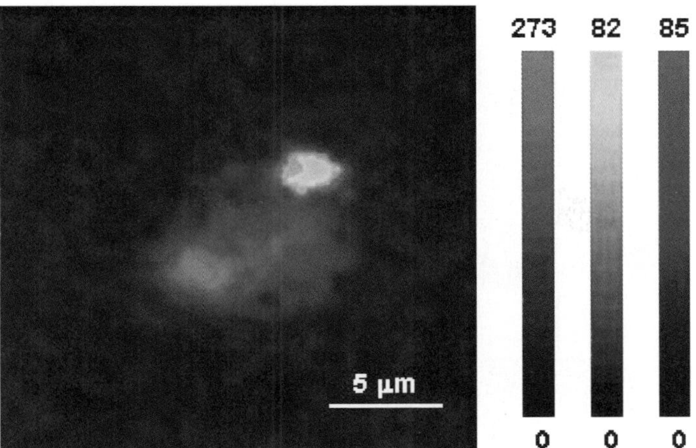

Fig. 4.20. This image was obtained from a macrophage cell cultured in a Cr solution environment (2 mg/L medium) for 30 min. The ranges of measured fluorescent intensities are from 0 to 273 photons/s for Ca, from 0 to 85 photons/s for Zn and from 0 to 82 photons/s for Cr. Each range is divided into 20 levels. Each level has assigned a shade of red, blue and green, respectively

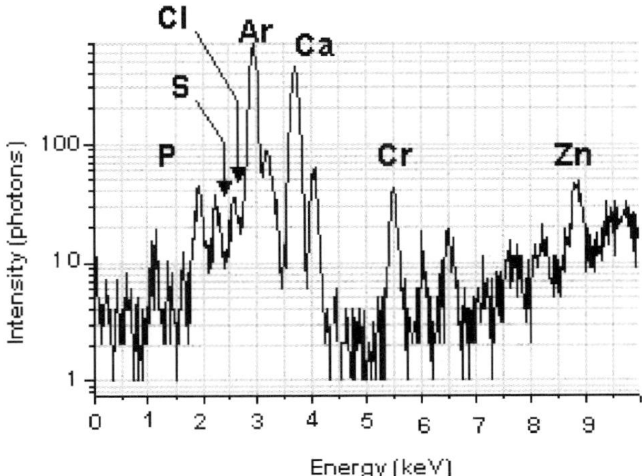

Fig. 4.21. The spectrum was obtained from the cell shown in Fig. 4.19a with the highest density of Ca. The measurement time 200 s and the incident X-ray energy was 14.3 keV

Effect of Cr Uptake on K Density

The value of the relative density of K within the cell shown in Fig. 4.19a was 0.690, suggesting that a part of the intracellular K was emitted out of the cell after exposing the cell to a Cr solution.

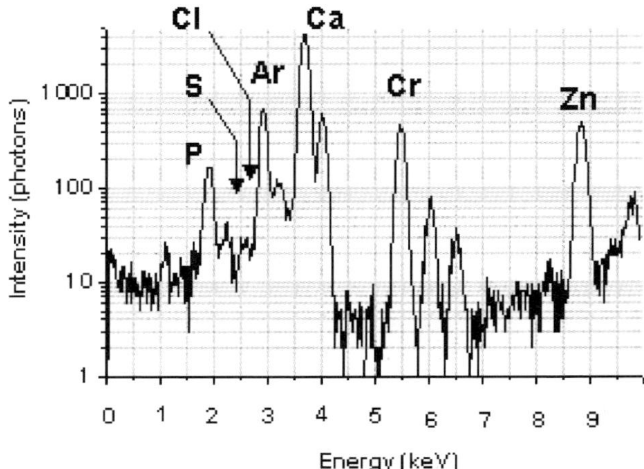

Fig. 4.22. The spectrum was obtained from the cell shown in Fig. 4.19b with the highest density of Ca and Cr. The measurement time 200 s and the incident X-ray energy was 14.3 keV

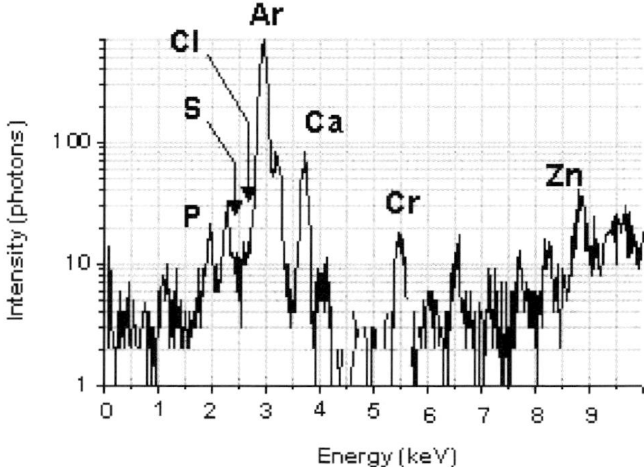

Fig. 4.23. The spectrum was obtained from the cell shown in Fig. 4.19c with the highest density of Ca and Cr. The measurement time 200 s and the incident X-ray energy was 14.3 keV

The value for the cell shown in Fig. 4.19b was 1.29, indicating that the density of K within the cell increased because of the rapid inflow of Cr into the cell.

For the cell shown in Fig. 4.19c, the value was almost 0. It appears that almost all of the intracellular K was emitted from the cell when the density of Cr remained low level after exposing the cell to a Cr solution.

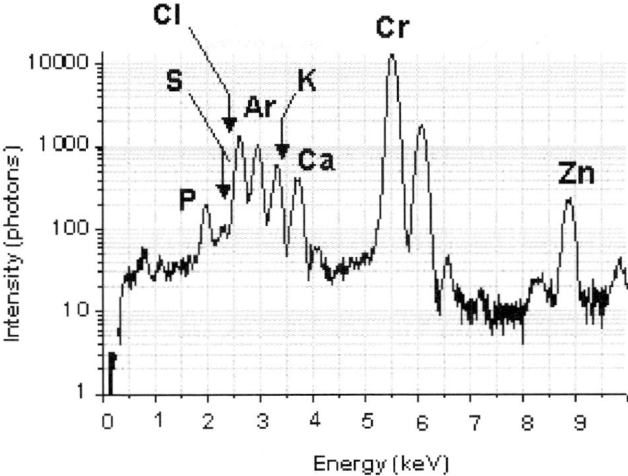

Fig. 4.24. The spectrum was obtained from the cell shown in Fig. 4.19d with the highest density of Ca. The measurement time 200 s and the incident X-ray energy was 14.3 keV

For the cell shown in Fig. 4.19d, the value was 5.28, suggesting that when the density of Cr within the cell steadily increased, the density of the intracellular K increased also.

Here also, the density of K within the cell may be correlated with the value of the uptake of Cr into the cell. Furthermore, it is probable that the value of the re-uptake of K is related to the condition of the cell membrane affected by the injurious effect of Cr solution environment.

Effect of Cr Uptake on Zn Density

The value of the relative density of Zn within the cell shown in Fig. 4.19a was 0.837, indicating very little outflow of Zn from the cell. For the cell shown in Fig. 4.19b, the value was 7.78, suggesting that after the rapid inflow of Cr into the cell, the subsequent inflow of Zn into the cell occurred rapidly. Furthermore, the distribution of Zn was localized adjacent to the population of Cr. The value for the cell shown in Fig. 4.19c was 0.445, indicating that the rate of the reduction of Zn from inside the cell is smaller than that of P or K. For the cell shown in Fig. 4.19d the value was 2.56, suggesting that the subsequent inflow of Zn was induced by the slow uptake of Cr into the cell over a long time.

The outflow of Zn may be restricted to a certain range. It is possible that Zn is indispensable for the cell in order to dispose foreign bodies. Furthermore, it is probable that Zn performs central role in defense against the toxic effect of the foreign metal element because of the fact that the subsequent rapid inflow of Zn is induced by the rapid inflow of Cr into the cell.

Effect of Cr Uptake on Ca Density

The value of the relative density of Ca within the cell shown in Fig. 4.19a was 0.385. Although the exposure time was very short, a large amount of Ca was emitted out from the cell. The value for the cell shown in Fig. 4.19b was 3.60, indicating that the inflow of Ca into the cell was induced by the rapid uptake of Cr. When the uptake of Cr was very fast in a short time, the relative density of Ca exceeded 1, which is the value for the control cell.

For the cell shown in Fig. 4.19c the value was 0.0648, indicating a rapid reduction of Ca within the cell. This outflow of Ca is the largest compared to those of other elements that can be detected by SR-XRF. The value for the cell shown in Fig. 4.19d was 0.232. In this case, the uptake of Cr increased slowly over a long time. It appears that the re-uptake of Ca may have occurred after the outflow in the initial stage.

Summary

In this study, Kitamura and Ektessabi found that the uptake of Cr ions or complexes causes characteristic changes in the distributions and densities of the intracellular elements in macrophages according to the speed and quantity of the uptake of Cr into the cell. It is clear that P, K, Zn and Ca are closely related to the uptake of foreign Cr into the cell. They found a rapid Cr uptake in the case of the cell shown in Fig. 4.19b, which was interpreted as an evidence that a part of the Cr added in the culture medium may be gathered temporally in a localized area around the cell in spite of the fact that the Cr solution was stirred. Therefore, macrophage cells in a highly condensed environment of Cr solution may be unable to avoid internalizing the high density of Cr. The fluctuations in the densities of the elements in the cells may be permitted within a certain range to sustain normal conditions after the uptake of foreign bodies. Yet, if the rapid inflow of Cr occurs, the rapid and excessive inflow of some elements may be induced in order to counteract the toxicity of Cr. From the results of SR-XRF analysis, it can be concluded that Ca, P and Zn play important roles in defense against the toxicity of Cr.

When the macrophage cells are cultured in a Cr chloride solution environment, the inflow of Cr occurs slowly. In the process of the uptake of Cr, the depletion of some elements such as Ca, K, P, S and Zn are observed within the cells. When the uptake and accumulation of Cr increase within the cells, the intracellular densities of P and Zn increase more than those in the control cells. However, the subsequent inflows of Ca, P and Zn were rapidly caused by the uptake of a large quantity of Cr in a very short time. These elements may be immediately required to defend against the foreign metal element.

Exposure to V Chloride Solution

Elemental Distribution in Treated Macrophages

Treated macrophages were cultured in a V chloride solution environment with different exposure times to investigate the differences in uptake of V and also the variations in the distribution of the internalized elements and intracellular elements. The exposure times were 0.5, 12 and 24 h. For each cell, seven images showing the distributions of P, S, Cl, K, Ca, Cr and Zn were obtained with the exposure time as a variable. The images showing the elemental distributions of Ca, Zn and V for three typical cases are shown in Fig. 4.25a–c. The images for Ca, Zn and V shown in Fig. 4.25a–c were from cells cultured in a V chloride solutions of 0.04 g/L medium for 0.5, 12 and 24 h, respectively. These images are 1 μm resolution, and the ranges of densities of Ca, V and Zn are each divided into twenty levels assigned to shades of green, red and blue respectively, as shown in Fig. 4.25a–c. In Fig. 4.25a_5 and b_5, the images of Ca and Cr overlap, and in Fig. 4.25c_4, the images of Ca and Zn

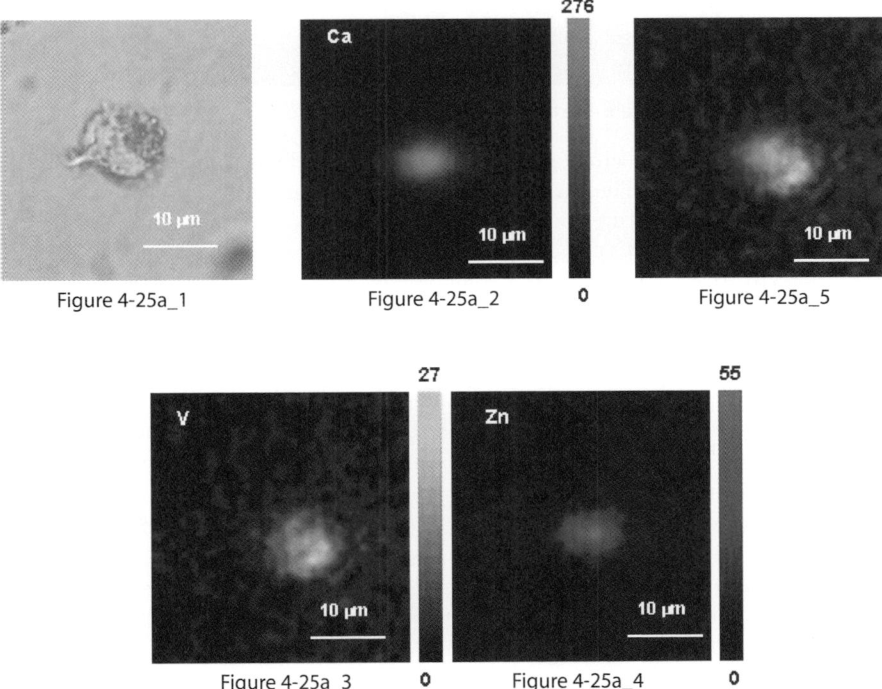

Figure 4-25a_1   Figure 4-25a_2   Figure 4-25a_5

Figure 4-25a_3   Figure 4-25a_4

Fig. 4.25a_1–a_5. A microscopic photograph of a dried macrophage dried on PET film is shown in (**a_1**). The elemental distribution of Ca, V and Zn within a single *treated* macrophage cultured in 0.04 g/L V solution for 0.5 h are shown in (**a_2**), (**a_3**) and (**a_4**), respectively. These images are matrices of 42 × 42 pixels of 1 μm resolution. In (**a_5**), the images of Ca and V are overlapped

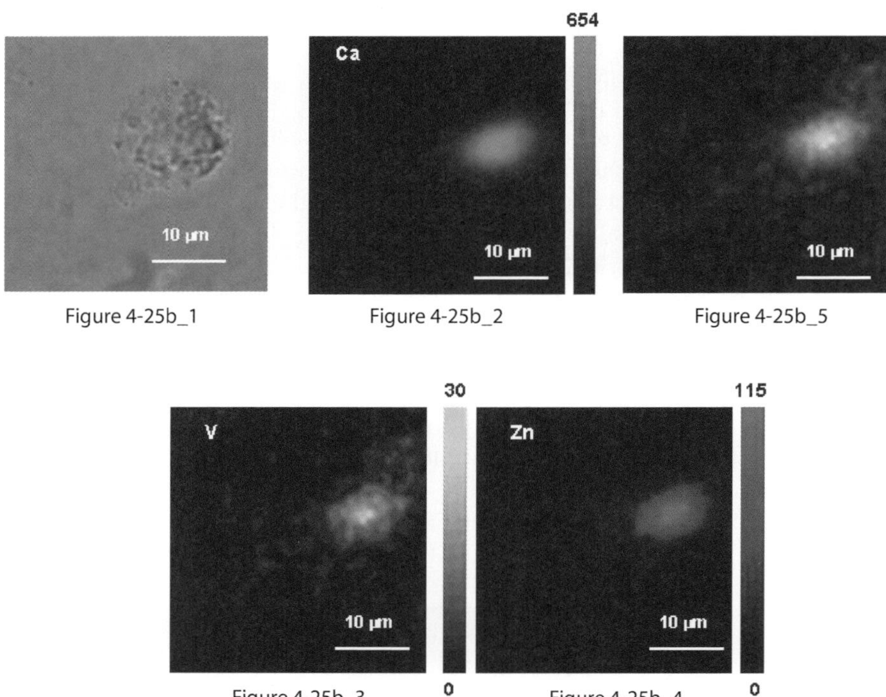

Figure 4-25b_1   Figure 4-25b_2   Figure 4-25b_5

Figure 4-25b_3   Figure 4-25b_4

Fig. 4.25b_1–b_5. A microscopic photograph of a dried macrophage dried on PET film is shown in (**b_1**). The elemental distribution of Ca, V and Zn within a single *treated* macrophage cultured in 0.04 g/L V solution for 12 h are shown in (**b_2**), (**b_3**) and (**b_4**), respectively. These images are matrices of 42 × 42 pixels of 1 μm resolution. In (**b_5**), the images of Ca and V are overlapped

overlap as well. The overlapping areas are expressed by their superimposed colors.

In the image shown in Fig. 4.25a, measured in 5 s, the maximum counts of the fluorescent X-ray of V, Ca, and Zn are 27, 276 and 55 counts respectively. The distribution patterns shown in Fig. 4.25a are typical images observed for the early stage of the uptake of V. The distribution patterns of Ca and Zn are almost identical with those in the control cell, but with higher densities. When the cells are cultured in a V solution environment, the uptake of V may be induced in a very short time.

In the image shown in Fig. 4.25b, the maximum X-ray intensities of V, Ca, and Zn are 30, 654 and 115 counts respectively, measured in 5 s. These are typical images seen at the state when the densities of Ca and Zn within the cell are increasing. The distribution patterns of Ca and Zn are almost identical with those in the cell shown in Fig. 4.25a. The quantity of V within the cell decreased compared to the early stage of the uptake of V.

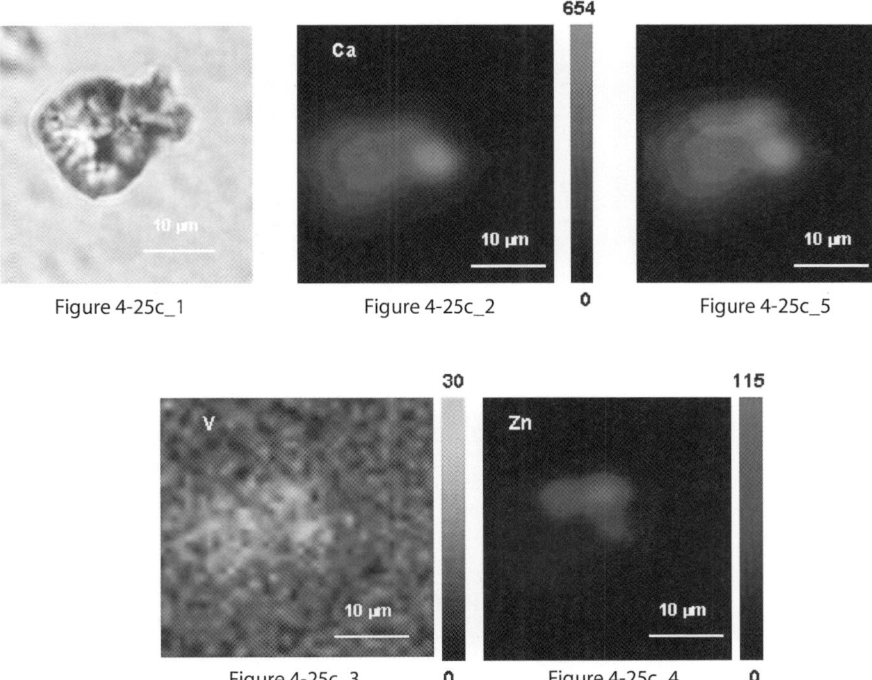

Figure 4-25c_1   Figure 4-25c_2   Figure 4-25c_5

Figure 4-25c_3   Figure 4-25c_4

Fig. 4.25c_1–c_5. A microscopic photograph of a dried macrophage dried on PET film is shown in (**c_1**). The elemental distribution of Ca, V and Zn within a single *treated* macrophage cultured in 0.04 g/L V solution for 24 h are shown in (**c_2**), (**c_3**) and (**c_4**), respectively. These images are matrices of 41×41 pixels of 1 μm resolution. In (**c_5**), the images of Ca and Zn are overlapped

In the image shown in Fig. 4.25c, the intensities for V, Ca, and Zn varied from 40 to 115, 11 to 4216 and 28 to 1164 counts respectively, also measured for 5 s. These were cultured in V solution environment for a long time. The distribution patterns of Ca and Zn are separated and adjacent each other. However, the quantity of V within the cell is almost zero.

Effect of V Uptake on P Density

The spectrum in Fig. 4.26 was obtained from the cell shown in Fig. 4.25a with the highest density of Ca. The cell was cultured in V chloride solution (0.04 g/L) for 0.5 h. The value of the relative density (defined as previously as the ratio of the particular element in the *treated* to that in the *untreated* cells) of P was 1.227. For other similarly cultured cells, the value of the relative density of P varied widely. It was anticipated that the density of P and V within the cell may be closely related, however such was not the case. The rapid inflow of V may not induce similarly rapid increase of P within the cell.

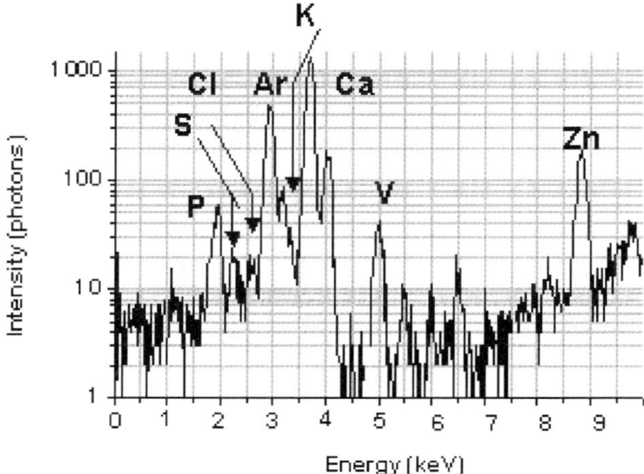

Fig. 4.26. The spectrum was obtained from the cell shown in Fig. 4.25a with the highest density of Ca. The measurement time 200 s and the incident X-ray energy was 14.3 keV

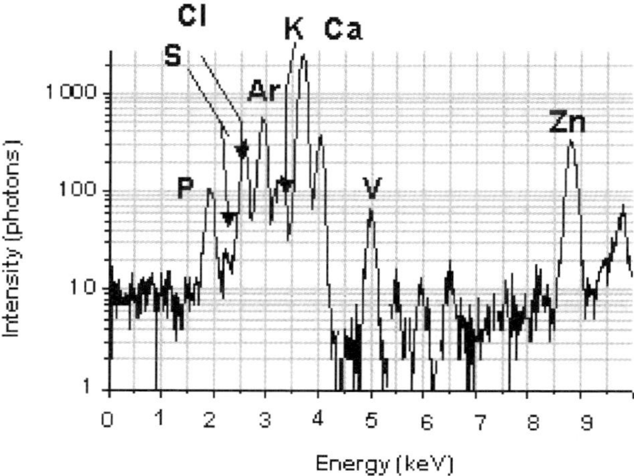

Fig. 4.27. The spectrum was obtained from the cell shown in Fig. 4.25b with the highest density of Ca. The measurement time 200 s and the incident X-ray energy was 14.9 keV

The spectrum in Fig. 4.27 was obtained from the cell shown in Fig. 4.25b with the highest densities of Ca. This cell was cultured in the same V chloride solution (0.04 g/L) for 12 h. The value of the relative density of P was 2.1287.

The spectra shown in Fig. 4.28 and 4.29 were obtained from the cell shown in Fig. 4.25c with the highest densities of Zn and Ca. The cell was cultured in a V chloride solution (0.04 g/L) for 24 h, and the value of the relative density of P was 0.5709. Almost nothing of the initially internalized V remained

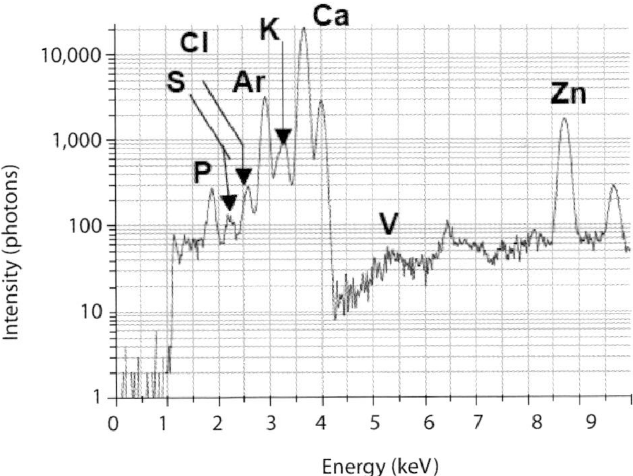

Fig. 4.28. The spectrum was obtained from the cell shown in Fig. 4.25c with the highest density of Ca. The measurement time 200 s and the incident X-ray energy was 14.3 keV

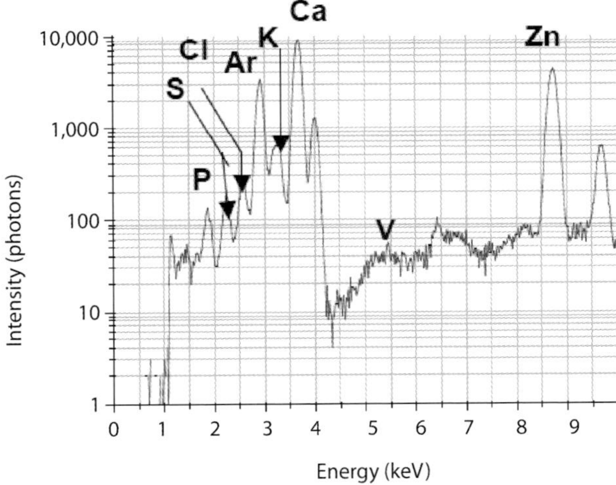

Fig. 4.29. The spectrum was obtained from the cell shown in Fig. 4.25c with the highest density of Zn. The measurement time 200 s and the incident X-ray energy was 14.3 keV

within the cell. It appears that the depletion of V may induce the outflow of P from inside the cell.

Effect of V Uptake on K Density

The value of the relative density of K within the cell shown in Fig. 4.25a was 0.7605, implying that part of the intracellular K may be emitted outside the cell after exposing the cell to a V solution.

For the cell shown in Fig. 4.25b the value was 2.034. The density of K within the cell increased because of the rapid inflow of V into the cell.

For the cell shown in Fig. 4.25c the value was 1.898. It is probable that the value of the re-uptake of K is related to the condition of the cell membrane affected by the injurious effect of V solution environment.

Effect of V Uptake on Zn Density

The value of the relative density of Zn within the cell shown in Fig. 4.25a was 4.587. Apparently the rapid inflow of V into the cell was accompanied by rapid inflow of Zn. Furthermore, the distribution of Zn was localized adjacent to that of Ca. The intracellular V induced much larger uptake of Zn into the cell compared to the case of Cr chloride solution case.

The value of the relative density of Zn within the cells shown in Fig. 4.25b and 4.25c were 7.996 and 16.44, respectively. The depletion of P was observed when the outflow of the internalized V occurred from inside the cell. However, for Zn the density within the cell remained at a high level in spite of the depletion of V.

It is probable that Zn performs a central role defending against the toxic effect of the foreign metal element because of the fact that subsequent rapid inflow of Zn is induced in both cases where rapid inflows of Cr and V occur into the cell. After the outflow of V from inside the cell, the population of Zn remained within the cell. The inflow of Zn into the cell may be required as a defense against the toxic effect of V, but the accumulation of the high density of Zn after the depletion of V may be toxic to the cell. It appears that macrophages with the excessive accumulation of Zn within their bodies cannot go back to the normal states. The excessive inflow of Zn beyond the permitted range for sustaining normal conditions may be injurious to the cells.

Effect of V Uptake on Ca Density

The value of the relative density of Ca within the cell shown in Fig. 4.25a was 2.017. The rapid inflow of Ca may be induced by the rapid inflow of V into the cell. In case of other cells exposed to the same solution for 0.5 h, the value of the relative density of Ca varied widely. It appears that a remarkable depletion of Ca within the cell was induced at the early stage of the uptake in both cases of Cr and V exposure. However, the inflow of Ca into the cell is induced by a small quantity of V internalized into the cell.

The values of the relative density of Ca within the cells shown in Fig. 4.25b and 4.25c were 2.946 and 4.289, respectively. The distribution pattern of Ca was localized and adjacent to the population of Zn. In case of other cells exposed to the same solution for 24 h, the value of the relative density of Ca varied widely. Depending on circumstances, although the relative density of Zn is much higher than 1 (within the control cell), the density of Ca within the cell is much less than in the control cell.

Summary

When the macrophage cells were cultured in Cr chloride and V chloride solution environments, there were some points in common concerning the changes of the elemental distribution patterns and the elemental density fluctuations within the cell. However, there are many different points that are very important in considering the differences of the toxic effect between Cr and V.

The outflow of V after initial uptake by the cell was observed in these experiments, posing a question. It is possible that the up-taking activity may be prevented by the unusual changes of the elemental densities within the cell. The first element that displays the rapid entry into the cell is Zn. Therefore, it can be concluded that the excessive uptake and accumulation of Zn within the cell prevent the cell from accumulating other elements. The maximum value of the relative density of Zn became 25.81 when the cells were cultured in V chloride solution environment for 24 h It was observed that the inflow of a large amount of Zn was caused by a small amount of V within the cell. The excessive accumulation of Zn within the cell is considered to induce not only the blocking of the uptake, but also an injurious effect to the cell. The localization and excessive accumulation of Ca within the cell was also observed in the area adjacent to the population of Zn. It is probable that the preservation of the high density of Ca within the cells may also be toxic.

4.2.4.2 Result of X-ray Absorption Fine Structure Analysis

4.2.4.2.1 Culture in Cr Chloride Solution Environment

In order to compare the chemical state of Cr before and after internalization, X-ray absorption fine structure (XANES) spectrometry was employed. The spectra obtained in the fluorescence mode were obtained from the different cells shown in Fig. 4.30 and 4.31. The corresponding cells were cultured in Cr chloride solution environments (0.04 and 0.004 g/L)

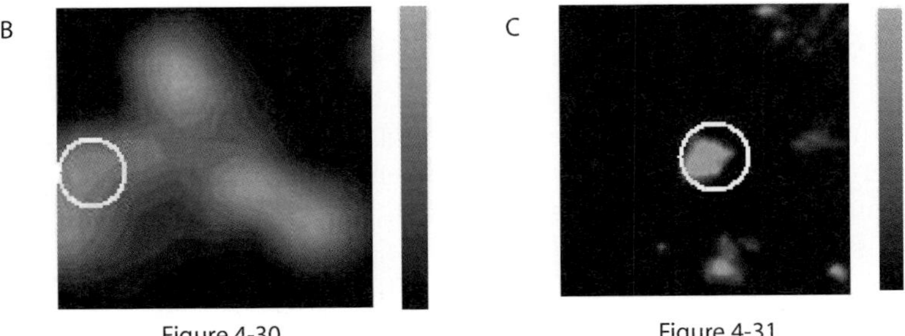

Figure 4-30   Figure 4-31

Fig. 4.30 and 4.31. Cells cultured for 24 h in CrCl$_3$ solution environment, 0.04 g/L for (**4.30**) and 0.004 g/L for (**4.31**)

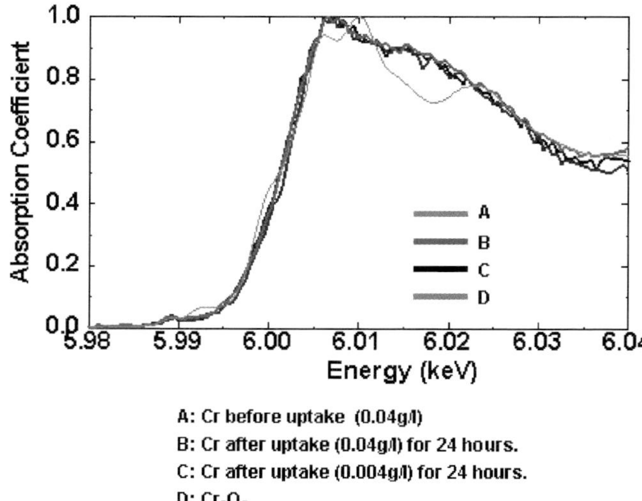

Fig. 4.32. The XANES was obtained from the different cells shown in Fig. 4.30 and Fig. 4.31. These spectra were obtained in fluorescence mode. The K-edge XANES spectra of Cr before the uptake and the reference spectrum of Cr_2O_3 are also shown

for 24 h. The K-edge XANES spectra of Cr within the cell under different conditions are shown in Fig. 4.32 together with the spectrum of Cr before internalization, as well as the reference spectrum, which is from Cr oxide (Cr_2O_3). The chemical structure around the element changes should be reflected in the X-ray absorption spectrum. It appears that the chemical states of Cr within the cells shown in Figs. 4.30 and 4.31 did not change because the three XANES spectra are virtually identical. While the distribution patterns and the densities of many intracellular elements changed significantly, it appears that the chemical structure around Cr atom within the cell does not change from the state before internalization.

4.2.4.2.2 Culture in Fe Chloride Solution Environment

The XANES spectrum obtained from the cell cultured in an Fe solution environment (0.04 g/L) for 24 h is shown in Fig. 4.33, together with the reference spectra from FeO and Fe_2O_3, and the spectrum of Fe before the internalization. Here, the absorption edge energy is defined at the point with half height to the maximum of the fluorescence yield. The change of the valence state induces shift of the absorption edge position. If the valence state changes from low to high, the absorption edge shifts to higher energy. Therefore, it is probable that the valence state of a portion of Fe internalized by the cell changes due to the chemical reaction within the cell. The pre-edge region of the spectrum shown in Fig. 4.33 is shown in Fig. 4.34. The pre-edge struc-

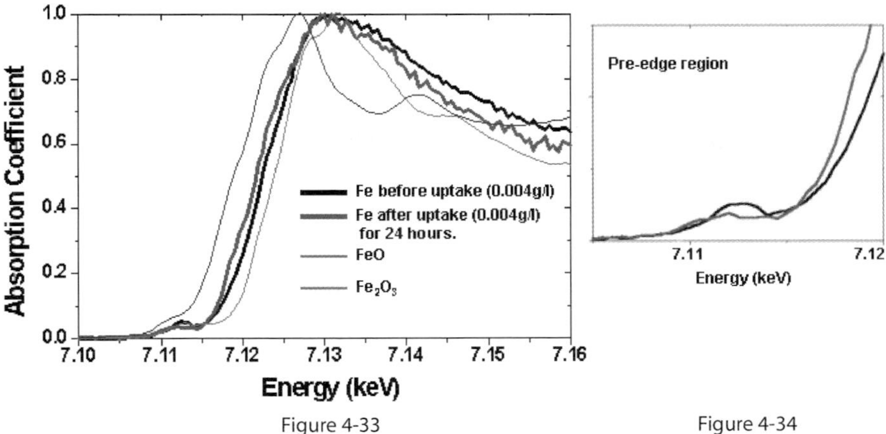

Fig. 4.33 and 4.34. XANES spectrum obtained from the cell cultured in 0.04 g/L Fe solution environment for 24 h is shown in (**4.33**). This spectrum was obtained in fluorescence mode. The reference spectra, which are FeO and Fe_2O_3, and the spectrum of Fe before the internalization are also shown. The pre-edge structure of the spectrum is shown in (**4.34**). It can be observed that there is a great difference between the two spectra in the pre-edge region

ture of X-ray absorption spectrum reflects the arrangement of the atoms surrounding the Fe. The data suggest that the coordination geometry of the site occupied by the atom surrounding the Fe may have changed within 24 h after the uptake of Fe into the cell leading to the transition of the valence state.

4.2.4.2.3 Summary

The change of the chemical state of the metal up-taken by the cell is possibly induced by the defensive reactions of macrophages against foreign metal elements or by the offensive reaction of the up-taken metal elements against macrophages. Confirmation of this possibility requires further investigations into the information of the chemical state of foreign metal elements internalized within the cell. For now, one can only conclude that the chemical structure of the metal elements taken up by the cell is closely related to the cytoxicity and the cell's defense mechanisms.

4.2.5 Consideration about the Interactions between Macrophages and Foreign Metal Elements

There are differences among the metal elements in the quantity internalized into macrophages. There are likely many factors in the interaction mechanisms between macrophages and foreign metal elements. For example, a mutual simple collision, the charged surface of macrophages and the mediation of the chemical substances affect the contact between macrophages and metals. However, the reason why macrophages show different take-up activities for different metal elements may be due to the interaction between the cells and metal elements after contact.

Macrophages have a built-in strategy against foreign metal elements in order to defend themselves, the other cells and tissues. The uptake of foreign metal elements is induced after contact with them. Macrophages may have quantitative restrictions of the level of internalized foreign metal elements within their bodies. Therefore, the quantities of metals within the cells are different. Yet, the interference of the up-taking activity by foreign metals may be closely related to their mutual interaction. For some metallic elements, the take up is prevented in the early stage, and its level within the cell drops due to the interference of the internalization. Why do the metal elements prevent macrophages from further activity after the internalization into the cells? Before considering this problem, it is necessary to consider the result of the excessive accumulation of Zn and Ca.

The density of Zn within the cell increased drastically when the rapid inflow of the foreign metal element was induced. The rapid inflow of Zn was observed in case of a V solution in spite of a little internalization. Why do macrophages need a lot of Zn within the cell when they are activated by foreign metal elements? It is probable that Zn plays an important role to promote the decomposition reaction of adenosine triphosphate (ATP). Macrophages need a lot of energy to realize the internalization and degradation of foreign particles and ions. ATP is produced in mitochondria, and a large amount of the energy used for various activities are generated by decomposing ATP. The generation of phosphoric acid is accompanied by decomposition process. By using EPMA imaging technique, one can observe that the internalized Cr composed macromolecules including P. It is considered that the decomposition of ATP is required to generate phosphoric acid for composing macromolecules with metal elements. Therefore, a large amount of Zn may be required in order to support the decomposition of ATP in the process of the internalization and digestion of metal elements.

Normally, macrophages perform the activity of internalization, digestion and excretion of foreign bodies. In this process, the homeostasis of the elemental densities in the cells is kept within a certain range. However, in case of the rapid inflow of the specific metal element into the cell, a large amount of Zn may be induced to enter into the cell accompanied by the destruction of homeostasis of elemental densities. The excessive accumulation of Zn may induce the subsequent production of reactive oxygen species (ROS). Free radicals are defined as atoms or molecules that contain one or more orbitals with a single unpaired electron. It is suggested that they tend to be highly reactive and capable of reacting with, and damaging, a variety of critical biologic molecules, including DNA, cytoskeletal proteins, and membrane lipids [21]. The mechanism of the increased free radical generation by Zn and the transport system for Zn is currently unknown, but it is suggested that Zn-induced production of ROS may be caused by the rapid entry of Zn into the cells [20, 22–24].

It appears that mitochondrial dysfunction may be induced by the free radicals produced by the rapid entry and excessive accumulation of Zn within the cell. The subsequent reduction in ATP synthesis and rise in intracellular Ca concentration are accompanied by the impairment of mitochondrial functions. It is suggested that such a rise of the intracellular Ca concentration can result in activation of Ca^{2+}-dependent proteases, lipases, and endonucleases, with resultant damage to the cytoskeleton, cell membrane, and DNA, respectively [25]. Here, when macrophages were cultured in a V solution environment, the densities of Ca and Zn displayed the rapid increase. Macrophages need Ca within them in order to internalize and digest foreign bodies. However, mitochondrial dysfunction can not dispose with a rise of Ca concentration induced by the uptake of foreign metal elements. Therefore, when the cell internalizes V, rapid entry of Zn into the cell and the subsequent mitochondrial dysfunction may be induced within the cell. Macrophages with mitochondrial dysfunctions can not avoid a rise of Ca density within the cell. If a rise of Ca density within the cell can not be relieved immediately, the further production of free radicals is induced and the subsequent mitochondrial dysfunction may be accelerated.

What is cytotoxicity? One can conclude that the strength of toxicity to macrophages equals the quantity of Zn entry per that of the internalized metal. If this value is very high, the subsequent process of the cell death may be induced rapidly in the early stage. However, the reason why each metal element has its own demand of Zn internalization for macrophages is still unknown. The chemical state of the metal element within the cell may be one of the most important factors related to Zn internalization. It is necessary to do further investigations into this problem. The mechanism of the cell death after interactions with up-taken metal elements are considered here and shown in Fig. 4.35.

In this study, the elemental distribution patterns and densities were mainly obtained by SR-XRF. By using SR source, the trace elements, such as Zn and other metals, can be detected and visualized as the elemental images of the cell. The detailed elemental images visualizing the sub-cell structure can be obtained by employing EPMA imaging technique. However, not all of the trace elements included within the cell can be detected by employing this technique because of the relatively poor detection limit ($> 0.1\%$). In order to obtain the trace and ultra-trace elemental distribution patterns of the sub-cell structure, a higher resolution for the SR microbeam is required. A resolution comparable to that obtaining in EPMA would enable better understanding of the roles of the ultra-trace elements in single cells.

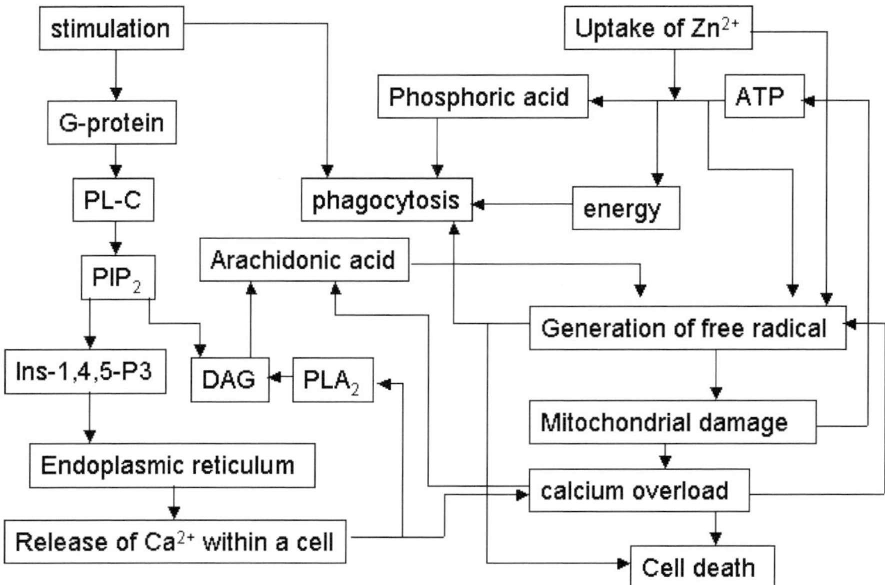

Fig. 4.35. The schematic drawing of the defensive mechanism and process of cell death

4.3 Elemental Images of Single Neurons by using SR-XRF

4.3.1 Introduction

The interactions and responses of neurons to transition metal elements are widely investigated because of the fact that transition metals can accept or donate single electrons to promote the free radical formation [26]. Iron or other transition metals such as copper are believed to induce neuronal injury by converting super-oxide ($O_2\bullet$) and hydrogen peroxide (H_2O_2) into highly reactive, toxic hydroxyl radicals ($OH\bullet$) in a sequence of reactions that is referred to cumulatively as the Haber–Weiss reaction [15]. It is suggested that neurodegenerative diseases are caused by these reactive oxygen species, and their excessive amount is generated through the reactions with transition metal elements, such as iron, zinc and copper. For example, aluminum is considered to be one of the most important sources related to Alzheimer's disease. Abreo et al. [27] displayed that the accumulation of aluminum in Neuro2a cells can result in the increased uptake of iron, inhibition of cell growth, and expression of NFT (neurofibrillary tangle) protein, partially mimicking the pathological hallmarks of Alzheimer's disease (AD), namely, loss of neurons and formation of NFTs [28]. On the other hand, copper is considered to possibly have a central role in several neurodegenerative disorders, including AD, Creutzfeldt–Jakob disease (CJD), and amyotrophic lateral sclerosis (ALS) [27]. Furthermore, in the case of ALS, it is suggested that the observed accumulation of calcium in motor neurons is closely related to the internalization of zinc into the cell [29, 30]. Increased iron levels within neurons have been observed in several neurological disorders, e.g., Parkinson's disease and Huntington's disease [31–34]. The storage of iron within ferritin may act as a protective mechanism, but heavily loaded ferritin may still produce free radicals. It is possible that the increased loading of ferritin in Parkinson's disease provides an environment that encourages free radical generation and hence neuronal damage [35].

It is probable that trace elements within neurons may display significant fluctuations when internalization and accumulation of certain transition metals are induced. In an experiment described in the following, the interactions between metals and neurons in vitro was investigated. A sample neuron, Neuro2a, was cultured in chromium oxide and vanadium chloride solution environments. After the cell culture, the cells were studied by SRXRF and EPMA-EDX analysis. It is anticipated that the ability to measure elements at ultra-trace levels may clarify their role within neurons affected by dosing metal elements. Especially, intracellular iron and calcium are considered to play important roles in order to control the fate of the cells against foreign metal elements. Iron is essential to tissue metabolism in brain and other organs. However, iron catalyzes lipid peroxidation and free radical production with results that could be especially destructive to a lipid-rich structure, such as the brain [19]. Normal brain iron may be toxic if it is released from its tightly controlled compartments and excessive accumulation of iron is induced within neurons [36].

Here, interactions between single neuron cells and metal elements are investigated via the measurement of the density fluctuation of the elements in the matrix and also of the localization of those elements in single cells, using samples that have been cultured in a metallic solution environment.

4.3.2 Procedures of Cell Culture and Morphological Observation

4.3.2.1 Neuron

The human brain – the control center that stores, computes, integrates, and transmits information – contains about 10^{12} neurons, each forming as many as a thousand connections with other neurons. The function of a neuron is to communicate information, which it does by two methods: electric signals processing and transmitting information within a cell. Chemical signals transmit information between cells, utilizing processes similar to those employed by other types of cells to signal each other. Information from the environment creates special problems because of the diverse types of signals that must be sensed – light, touch, pressure, sound, odorants, the stretching of muscles. Sensory neurons have specialized receptors that convert these stimuli into electric signals. These electric signals are then converted into chemical signals that are passed on to other cells – called interneurons – that convert the information back into electric signals.

4.3.2.2 Procedures of Cell Culture

Neuro2a was employed in this experiment. These cells were provided in American Type Culture Collections (ATCC). The culture medium was Eagle's Minimal Essential medium that is modified by ATCC to contain: 1.0 mM sodium pyruvate, 0.1 mM nonessential amino acids and 1.5 g/L sodium bicarbonate. 10% fetal bovine serum (FBS) was added into the culture medium. The cells in dimethyl sulfoxide (DMSO) and culture medium which were frozen in liquid nitrogen were defrosted in warm water (37 °C). After defrosting, the cells were pipetted into a sterilized tube with the culture medium (5ml). After the centrifugation of the tube, the cells were distributed into 550 mm diameter dishes with 5ml culture medium. The cells were cultured in an incubator fixed at 37 °C and 5% of concentration of carbon dioxide. When the number of cells increased and the dish was covered with a lot of cells, a part of the cells were moved into the new dish with the fresh culture medium.

After the cell division, axon and dendrites of each neuron spread in all directions and make networks with other neurons. The scanning electron microscope image of neurons making networks is shown in Fig. 4.36. In their cell bodies, there are many sub-cell structures. In order to make them clear, Giemsa's solution, methyl green solution, Mayer's hematoxylin solution and eosin Y ethanol solution (0.5%) were employed. The staining procedures were described in Sect. 4.2.2.3. The light microscopic photographs of neurons stained by these procedures are shown in Figs. 4.37, 4.38 and 4.39. The nucleus, the nucleoli in the nucleus, dendrites and axons are visible clearly in each figure.

The procedures for preparing the metallic solutions were as follows. The metal compounds utilized in this experiment were CrO_3 and VCl_3. These metal powders were enclosed in the bottle and were sterilized in an autoclave for 30 min. The metal powders were dissolved by the sterilized PBS solutions. The metallic solutions were further sterilized through the filter with 0.45 μm holes. If the cells were cultured in a metallic solution environment, the ratio of the culture medium and metallic solution was 100 to 1.

4.3.2.3 Morphologic Observation with Scanning Electron Microscope

Neurons have many dendrites and axons spreading from their bodies to create the networks with other neurons. The axons and thick dendrites can be seen by light microscope, but fine threads of dendrites cannot be observed by this method. In order to visualize the structures of the axons, dendrites and

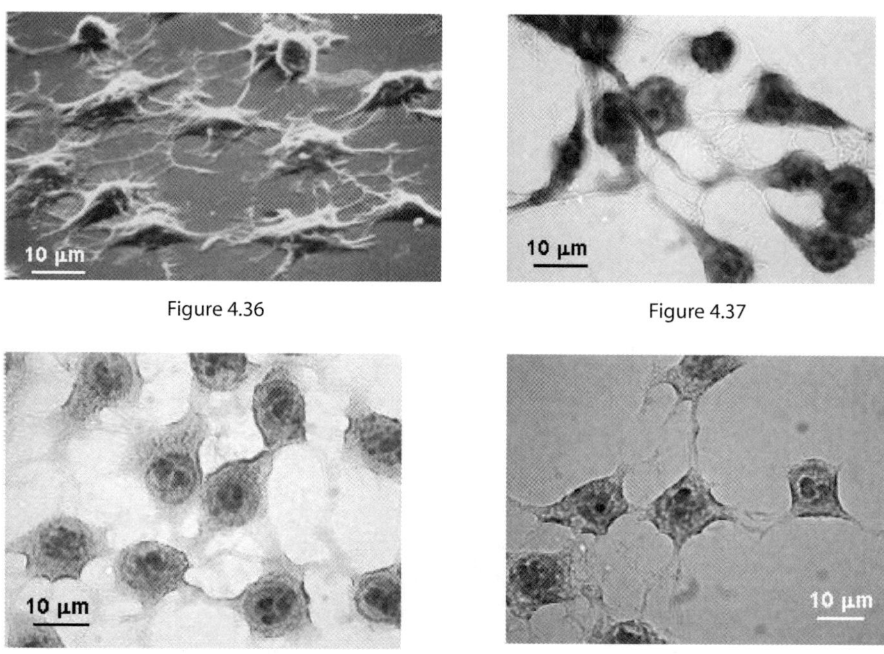

Figure 4.36

Figure 4.37

Figure 4.38

Figure 4.39

Fig. 4.36, 4.37, 4.38 and 4.39. The scanning microscope images of neurons making networks with other neurons (**4.36**). In order to make sub-cell structures clear, giemsa, hematoxylin-eosin and methyl green staining procedures are employed and shown (**4.37**), (**4.38**) and (**4.39**), respectively

cell bodies after exposing cells to a metallic solution environment, the SEM (scanning electron microscope) analysis was employed.

For SEM, silicon plates were employed as the substrate because of the good electro-conductivity. The silicon plates were immersed in ethanol solution and sterilized by the ultraviolet ray for 24 h Neuro2a were cultured on these silicon plates and fixed by a formalin solution for 24 h. After fixing, carbon was deposited on the silicon plate. Selection of the fixation solution is important to make the fine structures of the cells clear. In order to compare formalin fixation technique with other ones, the macrophage cells were fixed by ethanol, methanol and formalin solution, shown in Fig. 4.40, 4.41 and 4.42, respectively. When the cells were fixed by methanol or ethanol solution, the fine ruffle structures on the cell surfaces and fine threads of dendrites were not visible. With formalin, the fine structure of the cell bodies, dendrites and axons of the cells were visible distinctly as can be seen in Fig. 4.42.

The cell cultured under normal condition is shown in Fig. 4.43. The cell cultured in a vanadium chloride solution (0.04 g/L) for 1.5 h are shown in Fig. 4.44. The fine structures of the surface of the cell body and the many

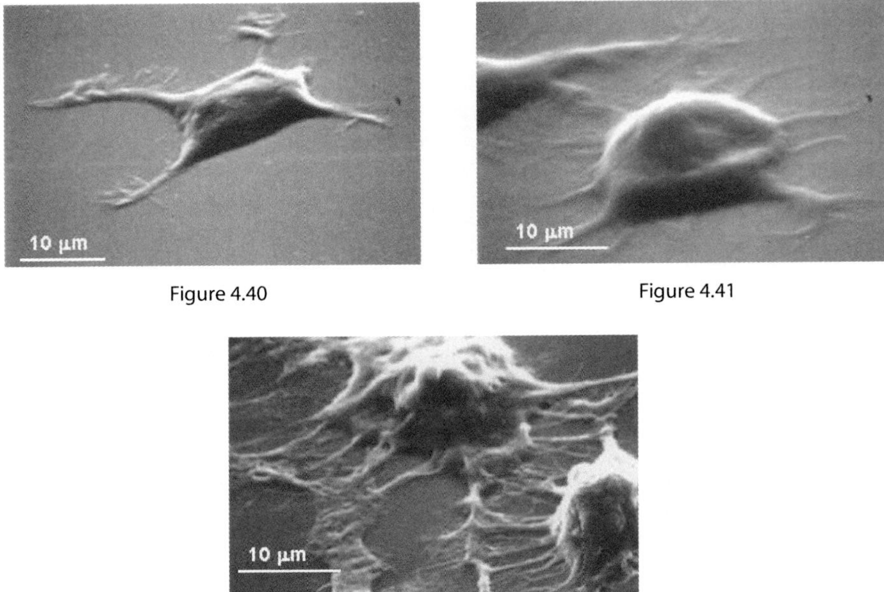

Figure 4.40

Figure 4.41

Figure 4.42

Fig. 4.40, 4.41 and 4.42. Macrophages fixed with ethanol, methanol and formalin solutions are shown in (**4.40**), (**4.41**) and (**4.42**), respectively

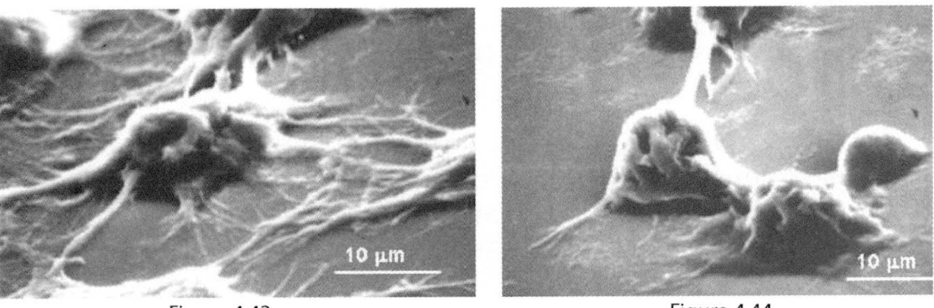

Figure 4.43

Figure 4.44

Fig. 4.43 and 4.44. Cells cultured under normal conditions are shown in (**4.43**). Cells cultured in a 0.04 g/L V chloride solution for 1.5 h are shown in (**4.44**)

fine threads of dendrites are clearly visualized in Fig. 4.43. However, the fine threads of dendrites are lost because of V solution environment, shown in Fig. 4.44. The lost of dendrites because of the V toxic effect is considered to be reasonable because the injured neuron with no dendrites cannot accept the signals from other cells and therefore the incorrect information can not spread around the injured cell. The question remains whether the V solution directly injures the fine threads of dendrites. There are many factors related to this phenomenon, but it is plausible that the effect of V solution may appear within the cell body first and a certain degenerative reaction may induce the degeneration of dendrites afterwards.

4.3.3 Sample Preparation and Experimental Methods

The X-ray fluorescence spectra and elemental images were mainly obtained at Beam Line 4A in High Energy Accelerate Institute using the X-ray beam emitted by 2.5 GeV storage ring "Photon Factory". The final beam size was 6 μm in diameter in this study. Fluorescent X-ray was detected by a SSD (Si(Li)) in the air. The incident beam energy was 14.2 keV.

The X-ray absorption spectra and some fluorescence spectra were measured at the Beam Line 39XU in Japan Synchrotron Radiation Research Institute using the X-ray beam emitted by the 8 GeV storage ring "SPring-8".

The cells (*treated* neurons) were cultured while being exposed to solutions containing chromium oxide, vanadium chloride and also cells without metal uptake (*untreated* neurons) were cultured under normal conditions. The Neuro2a mouse neurons were provided by the American Type Culture Collections. These neurons were cultured on the sterilized PET films in Eagle minimum essential medium supplemented with 10% fetal bovine serum. After exposure to metallic solutions, the cells were washed with culture medium and fixed in formalin solutions for 24 h.

4.3.4 Challenge for In Vivo and In Situ Measurement of Living Single Neurons

4.3.4.1 Objective

For the elemental analysis, it is possible to measure samples in the air and solutions by employing the SR source. For investigations into the cell functions, in vivo and in situ measurement are an ideal procedure. When the biological samples are measured by SR-XRF imaging technique, the fixed and thin sections of tissues or cells are normally employed. Here, the elemental distribution patterns are observed in vivo and in situ by SR-XRF imaging technique in order to obtain the real figure of the cell.

4.3.4.2 Procedure

In order realize in vivo and in situ measurement, it is necessary to perform the cell culture near the beam line. In the biological imaging center in SPring-8, it is possible to do cell culture for the measurement sample. At Kyoto University, neurons (Neuro2a) were cultured and frozen by liquid nitrogen. The frozen cells cooled in the dry ices were brought into the biological imaging center in SPring-8. These cells were defrosted and cultured on a thick PET film. The films were first sterilized using ethanol solution, washed with sterilized water and dried under ultraviolet rays. Two kinds of PET films,

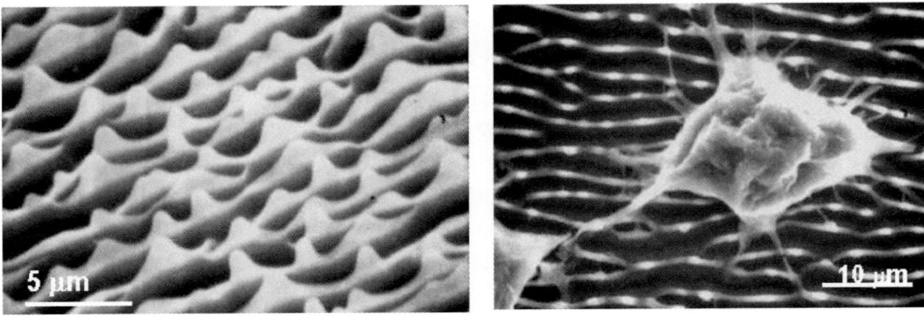

Figure 4-45   Figure 4-46

Fig. 4.45 and 4.46. Two kinds of PET films, normal and special ones, were prepared for the cell culture. Special PET films were processed by excimer laser irradiation in order to form the microstructures on the surfaces of them. The surface structures of the PET film after irradiation are shown in (**4.45**). The SEM photograph of the neuron cultured on the special PET film is shown in (**4.46**)

normal and special ones, were prepared for the cell culture. Special PET films were processed by the excimer laser irradiation in order to form the microstructures on their surfaces. The surface structures of the PET film after irradiation are shown in Fig. 4.45. It can be observed that there are many ruffles on the surface of the PET films. The SEM photograph of the neurons cultured on the special PET film is shown in Fig. 4.46. These microstructures on the surface of the PET film are designed to improve the adhesion power of the cells.

About a week later, the neurons on the surface of the PET films formed a tightly composed network with each other. Three kinds of samples were prepared for the measurement. They are the control neurons sample, and two treated neurons samples that were cultured in a chromium oxide solution ($0.04\,g/L$) for $4\,h$ on the surface of the normal PET film and on the surface of the special PET film. Before the measurement, the PET films were put on a thick acrylic container poured with the culture medium. The acrylic container has a hole at the bottom. The PET film is stretched over the hole and glued by adhesive agent. When the PET film with the cultured neurons is put on the acrylic container, the surface with neurons must be faced on the culture medium. After putting the film on the acrylic container, a thick acrylic cover with a big hole was pushed on the film. At the end, the neurons were confined between the two PET films filled with the culture medium. The thickness of the liquid phase is about 0.1 to 0.5 mm. After making the samples, they were brought into the beam line soon. The schematic drawing of procedures of making samples is shown in Fig. 4.47.

SR-XRF analysis was performed in the air. The strong adhesion power between the cells and film is required during the measurement because the film with the cultured neurons must be mounted vertically.

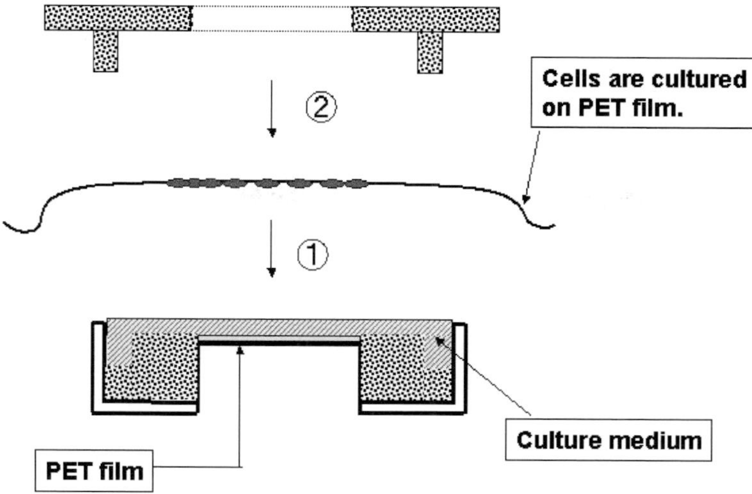

Fig. 4.47. Before in-vivo and in-situ measurement, the PET with living cells were put on the acryl container poured with the culture medium. The acryl container has a hole at the bottom of it. The PET film is stretched over the hole and glued by adhesive agent. When the PET film with the culture neurons is put on the acryl container, the surface with neurons must be faced on the culture medium. After putting the film on the acryl container, a thick acryl cover with a big hole was pushed on the film

4.3.4.3 Results

In order to obtain the elemental distribution pattern of the control cells cultured on the normal PET film, in vivo and in situ measurement was employed and the image is shown in Fig. 4.48a–d. When SR-XRF imaging technique was employed to the fixed cells, the distribution patterns of iron, calcium, potassium, phosphorus and sulfur were clearly observed within neurons. However, the distribution patterns shown in Fig. 4.48 are obscure. It is considered that the elements included in the culture medium shade off the gradation of the elemental distribution patterns within the cells.

The cells cultured in a chromium solution environment for 4 h on the normal and special PET films are shown in Fig. 4.48c and 4.48d, respectively. These images are also obscure.

Normally, SR-XRF imaging technique cannot detect the elemental distributions during watching the area under scanning. Therefore, the elemental distribution patterns obtained by SR-XRF must correspond to the photograph of the sample. In this system, if the elemental distributions do not display the distinct patterns, it is impossible to correspond the scanned images to the photograph of the sample. In order to realize the in vivo and in situ measurement, it is necessary to employ the real-time observation of

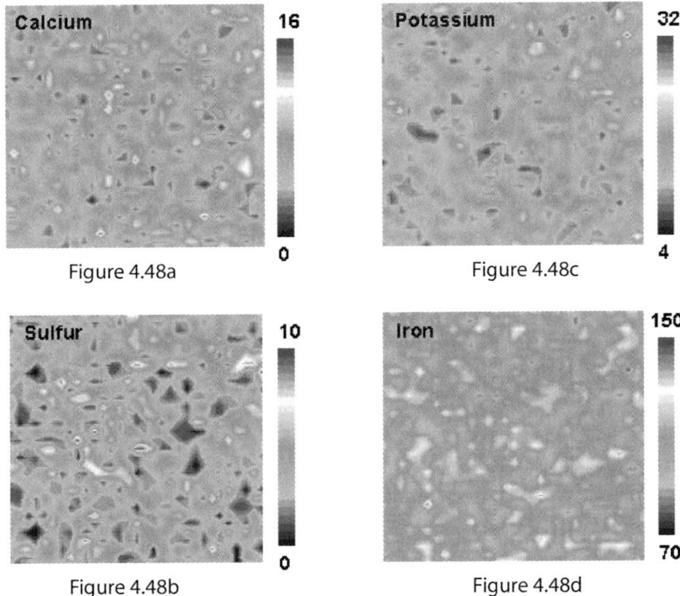

Fig. 4.48a–d. In order to obtain the elemental distribution pattern of the control cells cultured on the normal PET films, in-vivo and in-situ measurement were employed. The elemental images of Ca, K, S and Fe are shown in (**a**)–(**d**), respectively

the scanned area precisely. Furthermore, it must be considered that the contents of the culture medium are controlled to obtain the clear images of the cells.

4.3.4.4 Summary

By using synchrotron radiation source, in vivo and in situ measurement of the elemental images of the cell may be realized. It is considered that there are no methods except SR-XRF for obtaining the various elemental images within the living cell simultaneously. There are many phenomena that we have never observed within the cell. Especially, the instant and temporary reactions involved in the transfer of the many elements in the cell are considered to be important for the investigations into the cell functions, but it is impossible to obtain the real-time images of the intracellular elements by using the current procedures. It is probable that in vivo and in situ measurement of the elemental images of the cells with SR-XRF will be the essential method in order to approach the cell functions in spite of the incomplete technique, which needs to be further improved and developed.

4.3.5 Experimental Results

4.3.5.1 Elemental Images of Neurons

Elemental Distribution in Untreated Neurons

A SEM photograph of a neuron dried on PET film is shown in Fig. 4.49a. XRF spectra measured at points of the center of the cell body, the root of the axon and dendrite, and the axon are shown in Fig. 4.50a–d.

The elemental distributions of Ca, S and Fe within a single *untreated* neuron (the neuron that was cultured in non-metal environment) are shown in Fig. 4.49b–d, respectively. These images are matrices of 45×45 pixels of $1\,\mu m$ resolution. The range of the fluorescent X-ray intensity of Ca is from 0 to 15 counts, and this range is divided into twenty levels. Each level of the measured and interpolated points has been assigned to a corresponding shade of red, green and blue and plotted as shown in Fig. 4.49b. Similarly, Fig. 4.49c shows the distribution of S in the same cell and the range of the intensity of S is from 0 to 19 counts. Figure 4.49d shows a similar plot of

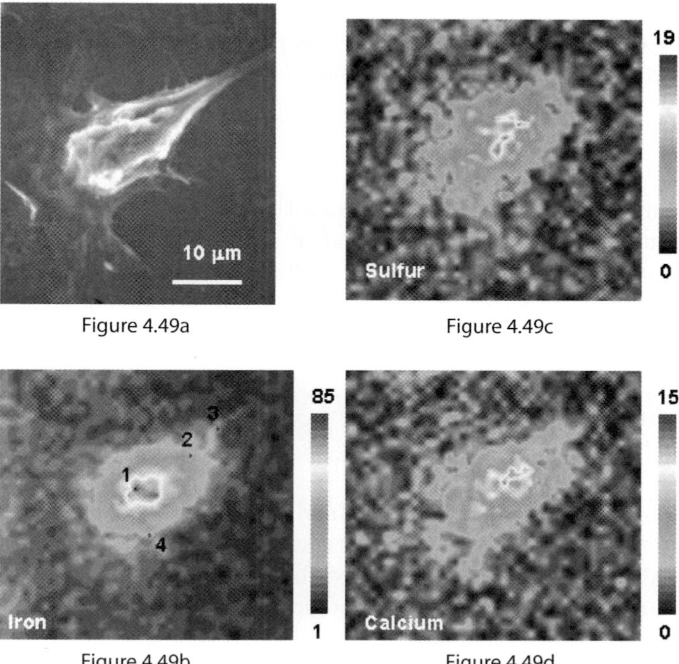

Figure 4.49a

Figure 4.49c

Figure 4.49b

Figure 4.49d

Fig. 4.49a–d. A SEM photograph of a neuron dried on a PET film is shown in (**a**). The elemental distribution of Ca, S and Fe within a single *untreated* neuron (neuron that was cultured in non-metallic environment) are shown in (**b**)–(**d**). In (**b**), the plotted points named No. 1 to 4 are defined as the center of cell body, axon hillock, axon and junction between axon and cell membrane

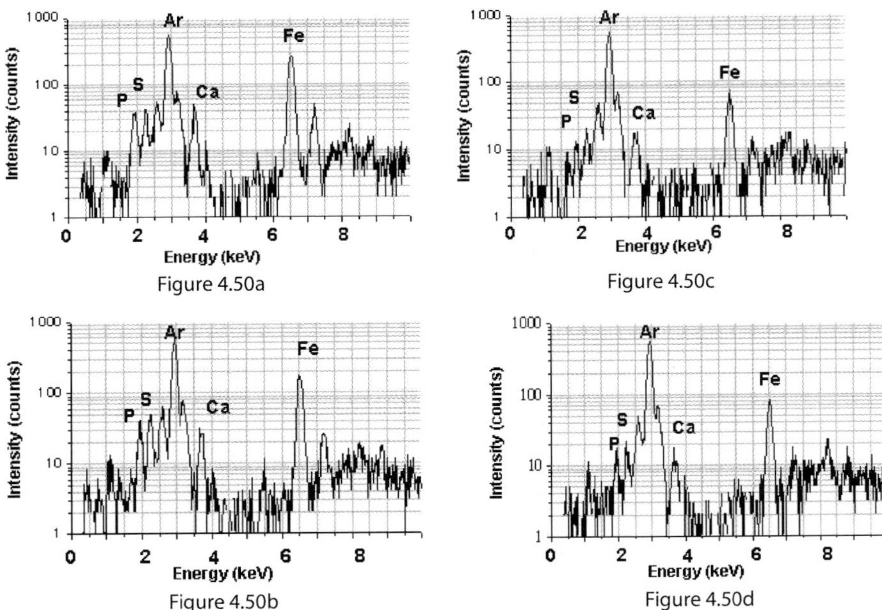

Figure 4.50a

Figure 4.50c

Figure 4.50b

Figure 4.50d

Fig. 4.50a–d. XRF spectra measured at the points of the center of the cell body, axon hillock and the junction between dentrites and cell membrane, are shown in **(a)**–**(d)**, respectively. The measurement time was 200 s and the excitation energy was 14.2 keV

the distribution of Fe in the same cell, with the range of the intensity from 1 to 85 counts. While the distribution of Ca, S and Fe in the neuron are almost identical as can be seen in Fig. 4.49b–d, the densities are completely different. These differences is evident in the case Fe, which is the strongest element detected in the X-ray energy range of 1 to 10 keV.

Exposure to Vanadium Chloride Solution

Elemental Distribution in Treated Neurons

Treated neurons were cultured in a vanadium chloride solution environment with different exposure times to investigate the difference in uptake of vanadium and also the variations in the distribution of the intracellular elements. The exposure times were 4 and 24 h. As for the cells cultured in the V chloride solution environment, four images for each cell – showing the distributions of P, S, Ca and Fe – were obtained with the exposure time as a variable. There was no evidence of internalization of V into the neurons.

The images of Fe, S and Ca within the neuron cultured in a vanadium solution (0.04 g/L) for 4 h are shown in Fig. 4.51b–d. These images are matrices of 45 × 45 pixels with a resolution of 1 μm. As before, the ranges of

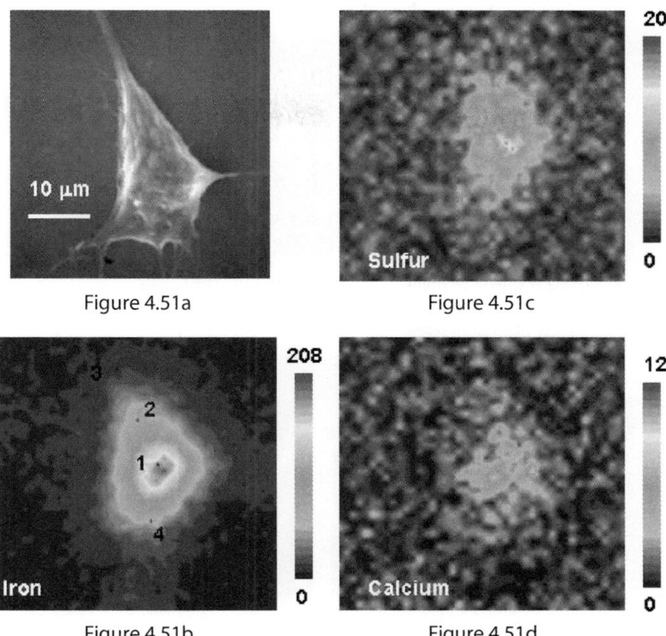

<table>
<tr><td>Figure 4.51a</td><td>Figure 4.51c</td></tr>
<tr><td>Figure 4.51b</td><td>Figure 4.51d</td></tr>
</table>

Fig. 4.51a–d. A SEM photograph of a neuron dried on a PET film is shown in (**a**). The elemental distribution of Ca, S and Fe within a single cell that was cultured in a 0.04 g/L V solution environment for 4 h are shown in (**b**)–(**d**), respectively. In (**b**), the plotted points named No. 1 to 4 are defined as the center of cell body, axon hillock, axon and junction between axon and cell membrane

density of Fe, S and Ca are each divided into 20 levels, and each level assigned to a shade of green, red and blue. The ranges of the florescent X-ray intensities of Fe, S and Ca are from 0 to 208, 0 to 20 and 0 to 12 counts respectively. The measurement time was 5 s. The experimental results shown in Fig. 4.51b–d reveal that these distributions have almost identical patterns as in the *untreated* cells. However, in Fig. 4.51c, the density of Ca within the cell decreased and the distribution patterns became obscure.

The images for Fe, S and Ca from neurons cultured in a V solution (0.04 g/L) for 24 h are shown in Fig. 4.52b–d. The matrices of these images are 45 × 45 pixels. As before, these images measured in 5 s, are of 1 μm resolution and the ranges of the intensities of Fe, S and Ca are from 0 to 86, 0 to 12, 0 to 8 counts, respectively. These results show that these distributions have almost identical patterns to those in *untreated* cells. Except for Fig. 4.52c, where the density of Ca within the cell decreased and the distribution pattern became obscure. In this case, the depletion of Ca from inside the cell is more remarkable than that in the case of the neurons cultured in a V solution for 4 h.

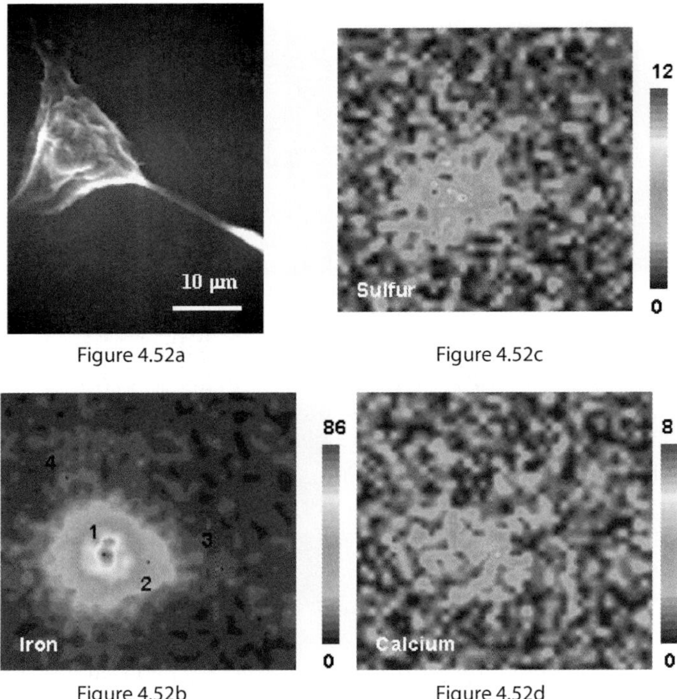

Fig. 4.52a–d. A SEM photograph of a neuron dried on a PET film is shown in (**a**). The elemental distribution of Ca, S and Fe within a single cell that was cultured in a 0.04 g/L V solution environment for 24 h are shown in (**b**)–(**d**), respectively. In (**b**), the plotted points named No. 1 to 4 are defined as the center of cell body, axon hillock, axon and junction between axon and cell membrane

Quantification of Density of Intracellular Elements as a Function of Time and Dose

For the quantification of the density of the intracellular elements, XRF spectra were obtained at the axon hillock (the junction of the dendrite and the cell membrane), the center of the cell body and the axon in each cell. The measurement time was 200 s. for each point. Each point is shown in Figs. 4.49, 4.51 and 4.52. In order to explain the increase or decrease in the density of the elements in relation to the effect of vanadium, it is necessary to invoke the parameter: "relative density" again (as in Sect. 4.2.4.1), defined as the ratio of the integrated value of fluorescent intensities of each element at the point in the *treated* neuron divided by that of the same element at the same point in the *untreated* one. A relative density of less than 1, implies that the density of that element in the *treated* cell is lower than that in the *untreated* one, conversely a value higher than 1 implies higher.

The intensities, normalized values and relative densities of elements at the points in the *untreated* and *treated* neurons are calculated and shown in Table 4.9 to 4.20. The results from the *untreated* cell are shown in Table 4.9 to 4.12. The results from the *treated* cell, which was cultured in a vanadium solution environment for 4 h, are shown in Table 4.13 to 4.16. The results from the *treated* cell, which was cultured in a V solution environment for 24 h, are shown in Table 4.17 to 4.20.

Cell Body

The cell body is the metabolic center of the cell. It contains the nucleus, which stores the genes of the cell, and the rough and smooth endoplasmic reticulum, which synthesizes the proteins of the cell.

Fe is the main element of the neuron detected in the energy range from 1 to 10 keV. In the control cell body, a large amount of Fe was observed and its peak intensity was detected at the center. After exposing the neuron to a V solution for 4 h, the density of Fe at the cell body center increased, which is shown in Table 4.13. On the other hand, the densities of P, S, Cl and Ca, decreased. The outflow of Ca was especially remarkable. After exposing the neuron to a V solution environment for 24 h, the density of Fe at the center increased even more and so did the depletion of Ca, compared to the case of 4 h exposure. The densities of P, S and Cl were still at a lower level than those in the control cell.

Axon Hillock

Electric disturbances generated in the dendrites or cell body spread to the axon hillock. The various depolarizations and hyperpolarizations move by passively spreading along the dendrite plasma membrane from the synapse to the cell body and then to the axon hillock. Whether a neuron generates an action potential in the axon hillock depends on the balance of the timing, amplitudes, and localization of all the various inputs it receives. Action potentials are generated whenever the membrane at the axon hillock is depolarized to a certain voltage called the threshold potential [37].

After exposing the neuron to vanadium solution for 4 h, the density of Fe at the axon hillock in the neuron increased, which is shown in Table 4.14. On the other hand, the densities of the other elements at the axon hillock, which are P, S, Cl, K and Ca, decreased after exposing to V. Especially the depletion of Ca from the neuron was the most remarkable. After exposing the neuron to a V solution environment for 24 h, the density of Fe at the axon hillock further increased. The density of P also increased more than that in the neuron cultured for 4 h, but that was still lower level than that in the control neuron. The densities of the other elements were also still at a lower level than those in the control cell.

Table 4.9

Center of cell body

Ar	1
P	0.0577874
S	0.065263
Cl	0.0937934
K	0.0495532
Ca	0.0673032
V	0
Cr	0
Fe	0.5865029
Cu	0
Zn	0

Table 4.10

Axon hillock

Ar	1
P	0.062354314
S	0.089460932
Cl	0.101415143
K	0.078982306
Ca	0.046679999
V	0
Cr	0
Fe	0.309780909
Cu	0
Zn	0

Table 4.11

Axon

Ar	1
P	0.019229759
S	0.026213471
Cl	0.087331254
K	0.048947706
Ca	0.031951731
V	0
Cr	0
Fe	0.115855142
Cu	0
Zn	0

Table 4.12

Junction

Ar	1
P	0.014106523
S	0.018333989
Cl	0.075479661
K	0.043152601
Ca	0.015510345
V	0
Cr	0
Fe	0.131122619
Cu	0
Zn	0

Table 4.13

Center of cell body

Element	Intensity	Relative
Ar	1	
P	0.05367	0.928749
S	0.051489	0.788946
Cl	0.0562	0.599189
K	0.07326	1.478411
Ca	0.020417	0.303359
V	0	
Cr	0	
Fe	1.427985	2.434474
Cu	0.021008	
Zn	0	

Table 4.14

Axon hillock

Element	Intensity	Relative
Ar	1	
P	0.016004	0.256662
S	0.027547	0.307922
Cl	0.051059	0.503465
K	0.016937	0.21444
Ca	0	0
V	0	
Cr	0	
Fe	0.52575	1.697167
Cu	0	
Zn	0	

Table 4.15

Axon

Element	Intensity	Relative
Ar	1	
P	0	0
S	0	0
Cl	0.039376	0.450881
K	0	0
Ca	0	0
V	0	
Cr	0	
Fe	0.100283	0.86559
Cu	0	
Zn	0	

Table 4.16

Junction

Element	Intensity	Relative
Ar	1	
P	0	0
S	0	0
Cl	0.036364	0.481773
K	0	0
Ca	0	0
V	0	
Cr	0	
Fe	0.109181	0.832663
Cu	0	
Zn	0	

Table 4.9–4.16. The intensities of the elements within the *untreated* cells are shown in (**4.9**) to (**4.12**). The intensities and relative densities of the elements in the *treated* cells with V solution environment for 4 h are shown in (**4.13**) to (**4.16**)

◄───

Table 4.17–4.20. Intensities and relative densities of the elements in the *treated* cells with V solution environment for 24 h

Table 4.17

Center of cell body

Element	Intensity	Relative
Ar	1	
P	0.051951	0.899002
S	0.050841	0.779017
Cl	0.057341	0.542053
K	0.032988	0.665709
Ca	0	0
V	0	
Cr	0	
Fe	0.96081	1.638202
Cu	0.029053	
Zn	0	

Table 4.18

Axon hillock

Element	Intensity	Relative
Ar	1	
P	0.039774	0.637873
S	0.036107	0.403606
Cl	0.038667	0.381275
K	0.026088	0.330302
Ca	0	0
V	0	
Cr	0	
Fe	0.617954	1.99481
Cu	0	
Zn	0	

Table 4.19

Axon

Element	Intensity	Relative
Ar	1	
P	0	0
S	0	0
Cl	0.060561	0.693464
K	0	0
Ca	0	0
V	0	
Cr	0	
Fe	0.102076	0.881066
Cu	0	
Zn	0	

Table 4.20

Junction

Element	Intensity	Relative
Ar	1	
P	0	0
S	0.008463	0.461604
Cl	0.03808	0.408057
K	0	0
Ca	0	0
V	0	
Cr	0	
Fe	0.098114	0.748262
Cu	0	
Zn	0	

Compared with the relative density at the center of the cell body, it can be observed that the tendencies of the change of the elemental densities are different. After 4 h, the density of Fe at the axon hillock increased, but the further increase was observed at the center of the cell body. On the other hand, the densities of P, S and Ca decreased at the center of the cell body, but the further decrease was observed at the axon hillock. Especially, the depletion of Ca from the axon hillock was remarkable. After 24 h, the density of Fe at the axon hillock further increased, but that at the center of the cell body decreased compared with that in the case of the cell cultured in a V solution for 4 h.

Junction of Dendrites and Cell Membrane

Most neurons have several dendrites. These branch out in tree-like fashion and serve as the main apparatus for receiving signals from other neurons. Dendrites are specialized to receive chemical signals form the axon termini of other neurons. Dendrites convert these signals into small electric impulses and transmit them to the cell body.

After exposing the neuron to a V solution for 4 h, the density of Fe at the junction of dendrites and cell membrane in the neuron decreased a little but this value was almost identical to that in the control cell, which is shown in Table 4.12 and 4.16. On the other hand, the densities of P, S and Ca decreased. After 24 h, the intracellular state at the junction of dendrites and cell membrane was almost identical to that after 4 h.

Axon

Most neurons have a single axon, whose diameter varies from a micrometer in certain nerves of the human brain to a millimeter in the giant fiber of the squid. Axons are specialized for the conduction of a particular type of electric impulse, called an action potential, away from the cell body.

After exposing the neuron to a V solution for 4 h, the density of Fe at the axon in the neuron was almost identical to that in the control cell, which is shown in Table 4.11 and 4.15. After 24 h, the density of Fe at this point had almost the same value. However, the densities of P, S and Ca decreased remarkably.

Summary

As a result of exposing the cell to a vanadium solution, the density of iron increased remarkably. Especially, the conspicuous increase of iron was observed at the axon hillock and center of the cell body. On the other hand, the depletion of calcium was observed in the cell. Internalization of vanadium into the cell was not observed. However, the SEM photographs revealed that many of the dendrites were lost after exposing the neurons to a vanadium solution environment. It is unclear whether vanadium solution directly injured the dendrites or whether the injurious effect to the cell body induced the subsequent degeneration of the dendrites. In either case, it is probable that iron is closely related to the defensive mechanism against foreign metal elements and the process of cell death.

Exposure to Chromium Oxide Solution

Elemental Distribution in Treated Neurons

Treated neurons were cultured in a chromium oxide (CrO_3) solution environment with different exposure times to investigate the differences in uptake of chromium and also the variations in the distribution of the intracellular elements. The exposure times were 0.5 and 4 h. As for the cells cultured in a Cr oxide solution environment, four images for each cell – showing the distributions of phosphorus, sulfur, chromium and iron – were obtained with the exposure time as a variable.

The images (elemental distributions) of P, S, Cr and Fe within the neuron cultured in chromium solution (0.04 g/L) for 0.5 h are shown in Fig. 4.53b, c, d and e. As previously, all these images are measured in 5 s, of 1 μm resolution and the ranges of density of P, S, Cr and Fe are each divided into twenty levels, each assigned to a corresponding shade (green, red and blue). The ranges of the fluorescent X-ray intensities of P, S, Cr and Fe are from 0 to 21, 0 to 17, 0 to 11 and 0 to 61 counts, respectively. The results show almost identical patterns to those from *untreated* cells.

The images for neurons cultured for 4 h are shown in Fig. 4.54b, c, d and e. The ranges of the florescent X-ray intensities of P, S, Cr and Fe are from 0 to 11, 0 to 10, 0 to 12 and 0 to 121 counts respectively. These results also reveal almost identical patterns to those from the *untreated* cells.

Relative Density of Intracellular Elements as a Function of Time and Dose

For the quantification of the density of the intracellular elements, the notion of relative density as defined previously will be used again. XRF spectra were obtained at the junction of dendrite and cell membrane, and the center of the cell body. In the case of V solution, XRF spectra were obtained at the axon hillock and axon. However, in this case, it was impossible to distinguish the axon from many of the dendrites. The measurement time was 200 s for each point. Each point is shown in Figs. 4.53 and 4.54. The relative densities at the axon hillock and axon were left out of consideration.

The intensities, normalized values and relative densities of elements at the points in the *untreated* and *treated* neurons are shown in Tables 4.21 to 4.24. The results from the *treated* cell, which was cultured in a Cr oxide solution environment for 0.5 h, are shown in Table 4.21 and 4.22, and for the 4 h in Tables 4.23 and 4.24.

Center of Cell Body

In the cell body of the control cell, a large amount of Fe was detected and its peak intensity was detected at the center of the cell body. After exposing the neuron to Cr oxide solution for 0.5 h, the density of Fe at the center of the cell body in the neuron decreased a little, as shown in Table 4.21. On

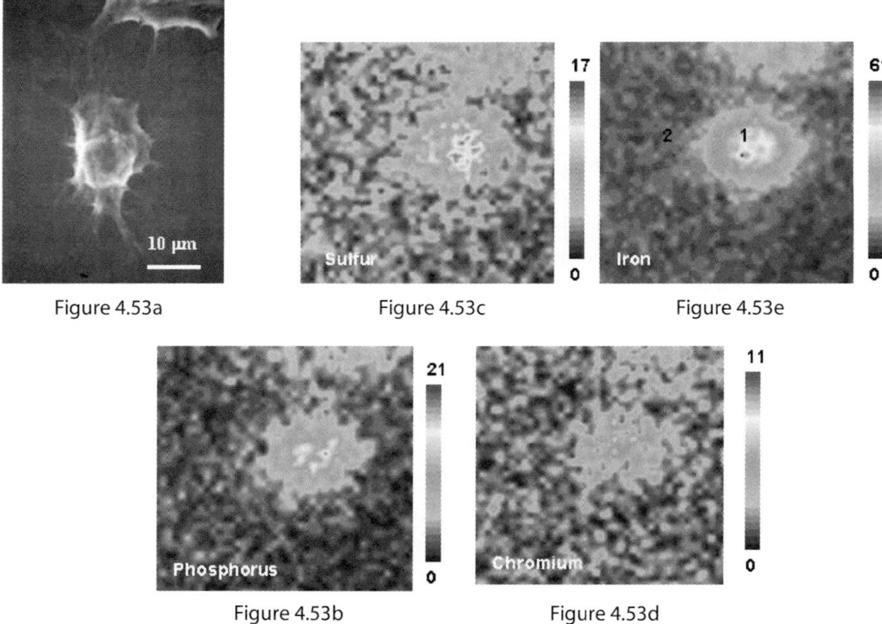

Figure 4.53a   Figure 4.53c   Figure 4.53e

Figure 4.53b   Figure 4.53d

Fig. 4.53a–e. A SEM photograph of a neuron dried on a PET film is shown in (**a**). The elemental distribution of P, S, Cr and Fe within a single cell that was cultured in a 0.04 g/L Cr solution environment for 0.5 h are shown in (**b**)–(**e**), respectively. These images are matrices of 43×43 pixels of 1 μm resolution. In (**e**), the plotted points named No. 1 and 2 are defined as the center of cell body and junction between axon or dentrites and cell membrane

the other hand, the densities of P, S, Cl and Ca were almost identical to those in the control cell. The densities of P and S increased a little, but they should not be considered to be significant changes because these values of fluorescence intensities were small. After exposing the neuron to a Cr oxide solution environment for 4 h, the density of Fe at the center of the cell body increased. The densities of P and S were almost identical to those in the control cell. On the other hand, the depletion of Ca was observed at this point.

Compared with the increase of the Fe density in the neuron cultured in a V solution environment, the increase of Fe is considered to be remarkable in this case.

Junction of Dendrites and Cell Membrane

For the 0.5 h exposure sample, the density of Fe at the junction of dendrites and cell membrane in the neuron was almost identical to that in the control cell, as shown in Table 4.22. The densities of P and S did not display any significant changes. The depletion of Ca was observed. For the 4 h sample, the density of Fe at the junction of dendrites and cell membrane increased

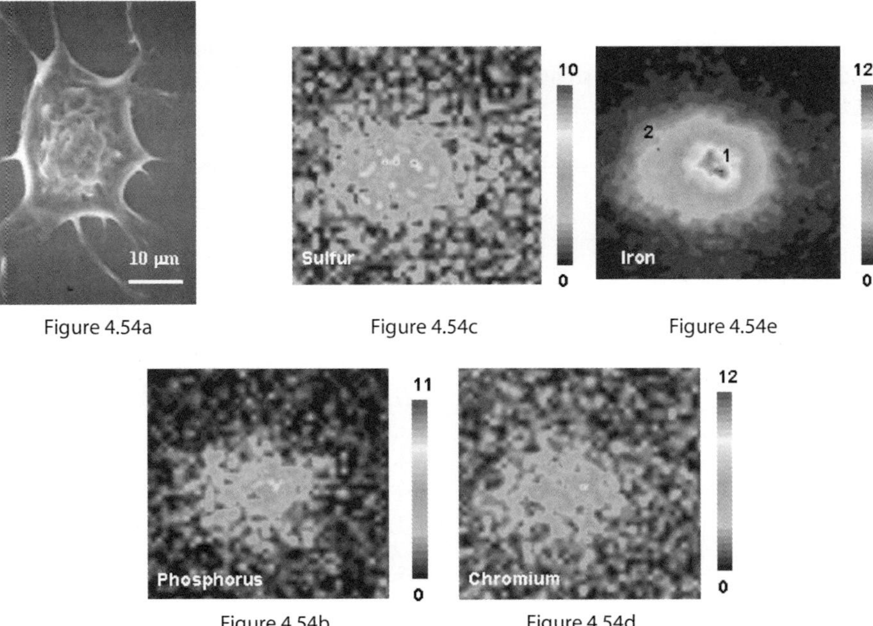

Figure 4.54a

Figure 4.54c

Figure 4.54e

Figure 4.54b

Figure 4.54d

Fig. 4.54a–e. A SEM photograph of a neuron dried on a PET film is shown in (**a**). The elemental distribution of P, S, Cr and Fe within a single cell that was cultured in a 0.04 g/L Cr solution environment for 4 h are shown in (**b**)–(**e**), respectively. These images are matrices of 45 × 45 pixels of 1 μm resolution. In (**e**), the plotted points named No. 1 and 2 are defined as the center of cell body and junction between axon or dentrites and cell membrane

remarkably. Furthermore, the densities of P, S and Ca also increased at this point. The relative density of Fe at the junction of dendrites and cell membrane is calculated by using the value of the control cell at the same point. Thus, it is plausible that the high density of Fe detected at the junction point between dendrites and cell membrane may be attributed to the redistribution of Fe.

Summary

When trypan blue dye exclusion test was performed to the macrophage cells cultured in a metallic solution environment, toxicity of chromium oxide (CrO_3) solution to the cell was the strongest of all metallic solutions. After exposing the neuron to a chromium oxide solution environment, the increase of the density of iron within the cell was relatively higher than that within the cell cultured in a vanadium solution. Therefore, it is probable that toxicity of the metal is related to the internalization of iron into the neuron. The internalization of vanadium by neurons was not observed, but that of chromium was observed. It is unclear why neurons did not internalize vanadium.

Table 4.21–4.24. The intensities and relative densities of the elements in the *treated* cells with Cr oxide solution environment for 0.5 h are shown in (**4.21**) and (**4.22**). The results of the treated cells with Cr oxide for 4 h are shown in (**4.23**) and (**4.24**)

Table 4.21

Center of cell body

Element	Intensity	Relative
Ar	1	
P	0.100647	1.741677
S	0.103085	1.579532
Cl	0.05455	0.581597
K	0	0
Ca	0.054138	0.80439
V	0	
Cr	0	
Fe	0.4531	0.772545
Cu	0	
Zn	0	

Table 4.22

Junction

Element	Intensity	Relative
Ar	1	
P	0.017523	1.242193
S	0.032155	1.753855
Cl	0.049477	0.655502
K	0.028245	0.654538
Ca	0	0.80439
V	0	
Cr	0	
Fe	0.134177	1.023294
Cu	0	
Zn	0	

Table 4.23

Center of cell body

Element	Intensity	Relative
Ar	1	
P	0.064836	1.121975
S	0.089172	1.366348
Cl	0.060065	0.640391
K	0.030943	0.66468
Ca	0.023493	0.349062
V	0	
Cr	0	
Fe	1.717469	2.928321
Cu	0	
Zn	0	

Table 4.24

Junction

Element	Intensity	Relative
Ar	1	
P	0.037062	2.627299
S	0.042529	2.319692
Cl	0.063234	0.837763
K	0.013518	0.31326
Ca	0.016628	1.072062
V	0	
Cr	0	
Fe	0.710059	5.41523
Cu	0	
Zn	0	

4.3.5.2 Result of X-ray Absorption Fine Structure Analysis

In order to compare the chemical state of Fe before and after exposing the neurons to a Cr oxide solution, XANES spectrometry was carried out. The spectra shown in Fig. 4.57 were obtained from the different cells shown in Figs. 4.55 and 4.56. These spectra were obtained in fluorescence mode. The cells shown in Figs. 4.55 and 4.56 were cultured in a Cr oxide solution environment (0.04 g/L) for 4 h and in non-metallic solution (normal condition), respectively. The reference spectrum, which is iron oxide (Fe_2O_3), is also shown in Fig. 4.57. If the chemical structure around the element changes, it

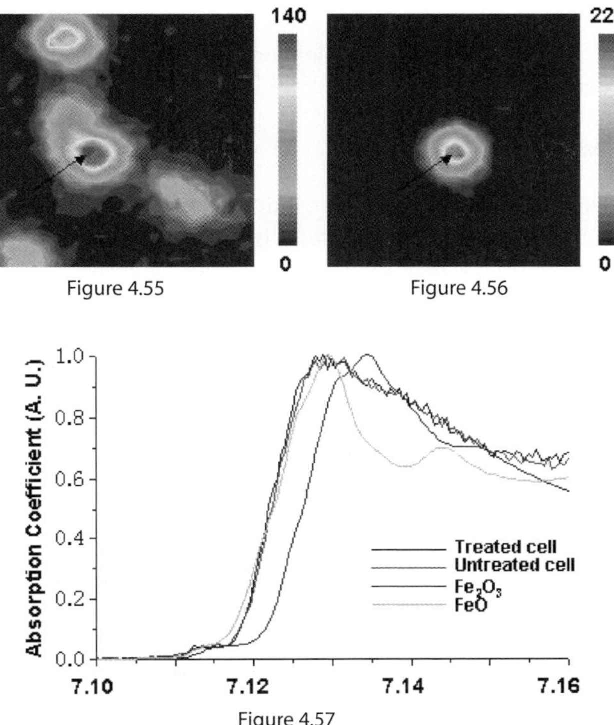

Fig. 4.55, 4.56 and 4.57. XANES spectra were obtained from the cells shown in (**4.55**) and (**4.56**), which were cultured in a 0.04 g/L Cr oxide solution environment and non-metallic solution for 4 h, respectively. XANES spectra of the cells and reference spectra of FeO and Fe$_2$O$_3$ are shown in (**4.57**)

is reflected in the structures of the X-ray absorption spectrum. The XANES spectra indicated that there was virtually no change of chemical state of Fe within the cells shown in Figs. 4.55 and 4.56. From the results of the elemental distribution images, it can be observed that the distribution patterns and the densities of many intracellular elements changed significantly. Yet, the chemical structures around Fe atoms were stable before and after internalization into the cell.

4.3.5.3 Results of EPMA Imaging

4.3.5.3.1 Experimental Set-up and Sample Preparation

The elemental images of single cells using EPMA-EDX were measured in the Department of Psychiatry in Nagoya University School of Medicine and on the JEOL-2010 electron microscope, equipped with a Norran Voyager energy dispersive X-ray micro-analyzer with a Si(Li) detector.

For EPMA measurement, the cells were immersed in glutaraldehyde and cacodylicacid. The thin section sample having 250 nm thickness was obtained from freeze-dried neurons.

4.3.5.3.2 Results

The neurons measured by EPMA were cultured in a V solution (0.04 g/L) environment for 4 h. The TEM photograph of the sample is shown in Fig. 4.58. The state of the sample was not good. It is unclear whether the toxic effect of V broke the cell membrane or whether the procedures of the sample preparation affected the membrane state. The elemental distribution patterns of P, S and V are shown in Figs. 4.59, 4.60 and 4.61, which were prepared similarly to the previous figures. The population of V within the neuron was located along the nuclear membrane. The density of P was also high in the same area. P and S displayed relatively uniform distribution patterns, but V was concentrated along the nuclear membrane of the neuron.

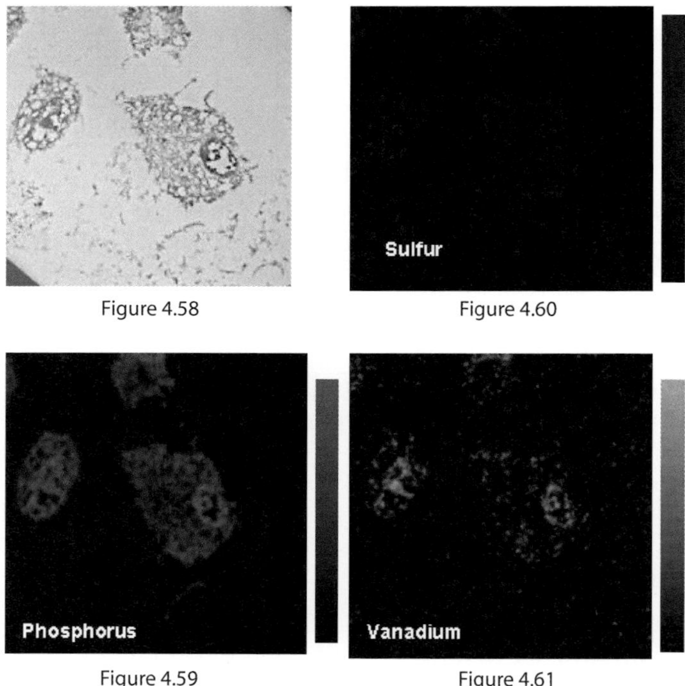

Figure 4.58   Figure 4.60

Figure 4.59   Figure 4.61

Fig. 4.58, 4.59, 4.60 and 4.61. The TEM photograph of the neurons cultured in 0.04 g/L V solution environment for 4 h is shown in (**4.58**). The elemental distribution patterns of P, S and V measured with EPMA are shown in (**4.59**)–(**4.61**), respectively

4.3.6 Discussion about the Interactions between Neurons and Foreign Metal Elements

After exposing neurons to vanadium chloride and chromium oxide solution environments, the immediate depletion of calcium was observed. Furthermore, the subsequent internalization of iron had occurred at the center of the cell body and axon hillock of the neuron. It appeared that chromium oxide may induce a larger amount of the internalization of iron than vanadium.

Iron is essential for the normal neurological function. It is said that iron uptake in most mammalian cells occurs via the transferrin cycle. Iron is normally transported in the plasma in the ferric state by transferrin [38]. Transferrin subsequently binds onto the transferrin receptor on the cell surface, which then undergoes endocytosis, generating endosomal vesicles within the cell. In a normal tissue, iron rarely exists as a free ion but rather is bound to a variety of active proteins including hemoglobin and myoglobin, transport proteins such as transferrin, and storage proteins such as ferritin [39]. Intracellular iron is usually tightly regulated, being bound by ferritin in an insoluble ferrihydrite core. When the intracellular ferritin iron was released, it was observed that it contributed to a free radical-induced cell damage in vivo [40]. Iron catalyzes lipid peroxidation and free radical production with results that could be especially destructive to a lipid-rich structure, such as the brain [19].

Why is the subsequent internalization of iron observed after exposing neurons to metallic solutions, such as vanadium and chromium? The SEM photograph of the neuron cultured in a chromium oxide solution environment displays that many of the thin branches of dendrites and axon were lost by a certain effect of chromium. It can be assumed that the first effect of the foreign metal to the neuron appears at the tip of the dendrites and axon. Proteins and membranes that are required for renewal of the axon are synthesized in the cell body. In the cell body, they are assembled into membranous vesicles or multi-protein particles, which are then transported along microtubules down the length of axon to the terminals. Axonal microtubules also are the tracks along which damaged membranes and organelles move up the axon toward the cell body. Biomedical processes in the brain that are dependent on iron include neurotransmitter synthesis, myelin production and maintenance, and basic cell functions such as energy production [41]. It is probable that the internalization of iron would be induced by the functional disorder of these transportation systems because of the injuries on the axon terminals.

It follows that the excessive accumulation of iron within the cell leads to an increased production of free radicals via the Fenton reaction [40]. Furthermore, free radicals cause the mitochondrial dysfunction and the subsequent excessive production of reactive oxygen species, such as O_2^-, H_2O_2 and $OH\bullet$. It has been suggested that the excessive formation of hydroxyl free radicals

(OH•) is cytotoxic and initiate lipid peroxidation and consequent cell damage [26].

In the case of the cell cultured in a chromium oxide for 4 h., the chemical state of iron within the cell did not change compared with that in the control cell. Therefore, it can be assumed that almost all of the internalized iron was converted to ferritin or other proteins at this density level. However, the more excessive accumulation of iron is considered to lead to production of reactive oxygen species.

It has been suggested that the production of nitric oxide is closely related to the accumulation of iron within the cell [42]. The amount of the free iron within the cell is assumed to be controlled by the production of nitric oxide. However, the excessive iron, which cannot be fixed as ferritin or other proteins, may form a large amount of O_2^- within the cell and result in the production of cytotoxic $ONOO^-$ and OH• after the interaction between nitric oxide and O_2^-. If the excessive accumulation of iron within the cell induces this process, further injurious reactions may be accelerated.

References

1. A.M. Ektessabi, M. Rokkum, C. Johansson, T. Albrektsson, L. Sennerby, H. Saisho, S. Honda, *J. Synchrotron Rad.*, **1998**, 5, 1136.
2. A.M. Ektessabi, J. Mouhyi, P. Louvette, L. Sennerby, *International Journal of PIXE*, **1997**, 7, 179.
3. A.M. Ektessabi, A. Wennerberg, *International Journal of PIXE*, **1995**, 5, 145.
4. J.C. Wang, W.D. Yu, H.S. Sandhu, F. Betts, S. Bhuta, R.B. Delamarter, *Spine*, **1999**, 24, 899.
5. J.Y. Wang, B.H. Wicklund, R.B. Gustilo, D.T. Tsukayama, *Biomaterials*, **1996**, 17, 2233.
6. A.S. Shanbhag, J.J. Jacobs, J. Black, J.O. Galante, T.T. Glant, *J. Biomed. Mater. Res.*, **1994**, 28, 81.
7. W.J. Kao, *Biomaterials*, **1999**, 20, 2213.
8. R. Geertz, H. Gulyas, G. Gercken, *Toxicology*, **1994**, 86, 13.
9. T. Pinheiro, M.L. Carvalho, C. Casaca, M.A. Barreiros, A.S. Cubha, P. Chevallier, *Nucl. Instr. and Meth. B*, **1999**, 158, 393.
10. P. Goegan, R. Vincent, P. Kumarathasan, J. Brook, *Toxicol. in vitro*, **1998**, 12, 25.
11. M. Radloff, M. Delling, G. Gercken, *Toxicol. Lett.*, **1998**, 96, 69.
12. G. Gercken, I. Berg, M. Dorger, T. Schluter, *Toxicol. Lett.*, **1996**, 88, 121.
13. J.C. Wataha, C.T. Hanks, Z. Sun, *Dental Materials*, **1995**, 11, 239.
14. S. Meresse, O. Steele-Mortimer, E. Moreno, M. Desjardins, B. Finlay, J.P. Gorvel, *Nature Cell Biology*, **1999**, 1, E183.
15. D. Granchi, G. Ciapetti, S. Stea, L. Savarino, F. Filippini, A. Sudanese, G. Zinghi, L. Montanaro, *Biomaterials*, **1999**, 20, 1079.
16. A. Aderem, D.M. Underhill, *Annu. Rev. Immunol.*, **1999**, 17, 593.
17. D.M. Underhill, A. Ozinsky, A.M. Hajjar, A. Stevens, C.B. Wilson, M. Bassetti, A. Aderem, *Nature*, **1999**, 401, 811.

18. J.A. Hunt, P.J. McLaughlin, B.F. Flanagan, *Biomaterials,* **1997**, 18, 1449.

19. A.M. Ektessabi, S. Yoshida, K. Takada, *X-ray Spectrometry,* **1999**, 28, 456.

20. F. Supek, L. Supekova, H. Nelson, N. Nelson, *J. Exp. Biol.,* **1997**, 200, 321.

21. C.E. Lewis, J. McGee, *"The Macrophage"*, IRL press, **1992**.

22. C.W. Olanow, G.W. Arendash, *Curr. Opin. Neurol.,* **1994**, 7, 548.

23. S.L. Sensi, H.Z. Yin, S.G. Carriedo, S.S. Rao, J.H. Weiss, *Neurobiology,* **1999**, 96, 2414.

24. K.M. Noh, Y.H. Kim, J.Y. Koh, *J. Neurochem.,* **1999**, 72, 1609.

25. S. Orrenius, M.J. Burkitt, G.E. Kass, J.M. Dypbukt, P. Nicotera, *Ann. Neurol.,* **1992**, 32, S33.

26. E. Kienzl, L. Puchinger, K. Jellinger, W. Linert, H. Stachelberger, R.F. Jameson, *J. Neurol. Sci.,* **1995**, 134, S69.

27. K. Abreo, F. Abreo, M.L. Sella, S. Jain, *J. Neurochem.,* **1999**, 72, 2059.

28. K. Zaman, H. Ryu, D. Hall, K. O'Donovan, K. Lin, M.P. Miller, J.C. Marquis, J.M. Baraban, G.L. Semenza, R.R. Ratan, *J. Neurosci.,* **1999**, 15, 9821.

29. W.Y. Ong, M.Q. Ren, J. Makjanic, T.M. Lim, F. Watt, *J. Neurochem.,* **1999**, 72, 1574.

30. P.F. Good, D.P. Perl, L.M. Bierer, J. Schmeidler, *Ann. Neurol.,* **1992**, 31, 286.

31. A.R. White, A.I. Bush, K. Beyreuther, C.L. Masters, R. Cappai, *J. Neurochem.,* **1999**, 72, 2092.

32. P.F. Good, C.W. Olanow, D.P. Perl, *Brain Res.,* **1992**, 593, 343.

33. K. Takada, A.M. Ektessabi, S. Yoshida, *AIP Conference Proceedings,* **1999**, 475, 452.

34. S. Yoshida, K. Takada, A.M. Ektessabi, *AIP Conference Proceedings,* **1999**, 475, 611.

35. P.D. Griffiths, B.R. Dobson, G.R. Jones, D.T. Clarke, *Brain,* **1999**, 122, 667.

36. K.D. Barron, *J. Neurol. Sci.,* **1995**, 134, 57.

37. H. Lodish, D. Baltimore, A. Berk, *"Molecular Cell Biology"*, third edition, W.H. Freeman & Co, **1995**.

38. J.R. Burdo, J. Martin, S.L. Menzies, K.G. Dolan, M.A. Romano, R.J. Fletcher, M.D. Garrick, L.M. Garrick, J.R. Connor, *Neuroscience,* **1999**, 93, 1189.

39. P. Ponka, *Blood,* **1997**, 89, 1.

40. K.L. Double, M. Maywald, M. Schmittel, P. Riederer, M. Gerlach, *J. Neurochem.,* **1998**, 70, 2492.

41. S. Chen, K.K. Sulik, *Pharmacol.,* **2000**, 294, 134.

42. G. Weiss, G. Werner-Felmayer, E.R. Werner, K. Grunewald, H. Wachter, M.W. Hentze, *J. Exp. Med.,* **1994**, 180, 969.

5

Investigation of Differentiation of Mouse ES Cells

5.1 Introduction

Embryonic stem (ES) cells are expected to lead to breakthroughs in the therapy for progressive neurodegenerative disorders such as Parkinson's disease, Alzheimer's disease and Huntington's disease [1–3]. ES cells are generally called pluripotent stem cells and are unique in that they have the capacity for unlimited self-renewal along with the ability to produce multiple different types of terminally differentiated descendants as shown in Fig. 5.1 [2]. The differentiation of ES cells can be controlled in vitro by choosing the configuration of culture conditions. The in vitro differentiation of mouse ES cell has been widely investigated and several methods to produce cardimyocytes, hematopoietic stem cells and endothelial cells are established [4–6]. To utilize ES cells for the therapy of neurodegenerative disorders, it is necessary to establish the method to culture dopaminergic neurons in vitro. The conditions required for neural cell induction had been unknown for a long time, but recently two groups identified their own methods using mouse ES cells [7,8].

McKay et al. cultured dopaminergic and serotonergic neurons in the presence of mitogen and specific signaling molecules and generated neuronal cells [9]. Another group lead by Sasai identified the substance that is generated from PA6 stromal cells and promotes neural differentiation of mouse ES cells. They named this substance stromal cell-derived inducing activity (SDIA) [10].

Although the procedure to induce neuronal differentiation is partially revealed, the details of the mechanism are unknown. Therefore it remains quite difficult to culture neurons efficiently for therapeutic application. Recently there have been several genetic studies to elucidate the mechanism of differentiation and organogenesis [11–14].

In this study, a new approach was taken to investigate the mechanism of differentiation by dissecting the change of distributions, concentrations and chemical states of intracellular trace elements. It is considered that trace metal elements and metalloproteins are deeply related to the orientation of

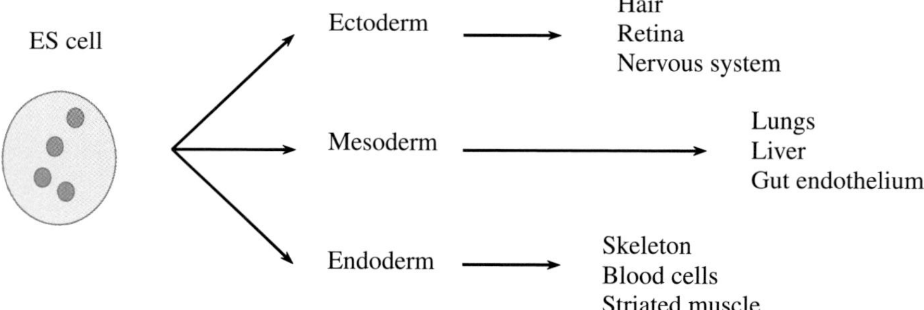

Fig. 5.1. Schematic drawing of pluripotency of mouse embryonic stem (ES) cells. ES cells are unique in that they have the capacity for unlimited self-renewal along with the ability to produce multiple different types of terminally differentiated descendants

differentiation as active centers as well as the neural cell death in neurodegenerative disorders [15].

The aim of this study is to analyze the distribution, concentration and chemical states of the trace elements in the process of differentiation of mouse ES cells, and to understand how they can be related to the differentiation. The investigations of the differentiation have never been carried out from the aspect of these elemental conditions at the cell level. From the experimental results, the unelucidated points, e.g., i) how the intracellular elements change in the process of the neuronal differentiation and ii) the optimal elemental conditions for neuronal differentiation are considered. The information obtained in this study will be valuable from the viewpoint of not only the neuroscience, but also basic biology about development of nervous system and evolution [16].

In this study, X-ray fluorescence (XRF) and X-ray near edge structure (XANES) analysis were applied to understand the multiple elemental conditions in different stages of differentiation efficiently and at the high sensitivity. Mouse ES cells form the colonies in the process of cell culture and each colony is in the different states of differentiation. Conventional chemical analysis methods enable simple quantification, but the information that is specific to each colony is lost in these methods because they require the fragmentation and solution of samples. XRF and XANES analysis does not require any pretreatment of samples to analyze the trace elements. These techniques are also nondestructive so it is possible to observe the progress of differentiation or the generation of neurotransmitters such as dopamine histochemically by immunostaining after the elemental analysis to the same samples. These features are a significant advantage that cannot be obtained in other techniques for studying the differentiation of mouse ES cells.

The study in this chapter can be divided into two experimental parts. In the first part of the study, the effect of differentiation to the intracellular trace elements was investigated. The change in concentrations and proportions of intracellular elements were investigated in the process of acquiring various functions and differentiating. The specific orientation of differentiation, such as the neuronal induction, was not performed in this experiment. After that, the neuronal differentiation was induced by the SDIA method that was suggested by Sasai et al. and the mechanism of neuronal development was considered. The chemical states of the transitional metal elements (Fe and Zn) are analyzed by XAFS technique in addition to elemental concentrations and distributions.

5.2 Investigation about the Effect of the Unoriented Differentiation

5.2.1 Cell Culture and Sample Preparation

Two groups first derived mouse ES cells from mouse embryos in 1981 [17,18]. Culturing ES cells technique are well established, and now it is the essential technique for the transfection and producing transgenic mice [1, 19]. Mouse ES cells are isolated from the inner cell mass (ICM) of post-conception mouse blastocysts. To maintain the pluripotency of the mouse ES cells, they have to be co-cultured with feeder layers of inactivated mouse primary embryonic fibroblasts. A feeder cell layer of fibroblasts prevents the differentiation of ES cells and makes them proliferate in undifferentiated states. The main aim of this study, however, is to consider the effect of the progress of the differentiation to the elemental distributions and concentrations. Therefore the feeder layer of fibroblasts were omitted to promote the differentiation into a variety of cell types and elicit the effect of differentiation.

The mouse ES cells (129/Sv) were purchased from Cell & Molecular Technologies, Inc. and the passage number (the age of cell line) was 15 in the beginning of the cell culture. The differentiation proceeds according to the increase of the passage number. The optical microscopic photographs of the cultured mouse ES cells whose passage numbers were 15, 16 and 17 are shown in Fig. 5.2.

Figure 5.2a shows the cells at the time of initial plating to gelatin-treated dishes and the passage number was 15. It can be seen that isolated cells adhered to the bottom of the culture dish. After 3 days the cells formed colonies as seen in b. Once the dishes are crowded and the colonies are large, the colonies are detached from the dish with Trypsin/EDTA, broken up into single cells and passaged into other gelatin-coated dishes. Figure 5.2c,d shows the photographs at the passage number 16 and 17, respectively. The morphological change of the ES cells of getting flat and dark can be observed in accordance with the repeated passages and it is the typical characteristic of the progress of the differentiation. The detailed culture procedure and the reagents for mouse ES cells are described in appendix 5.A.

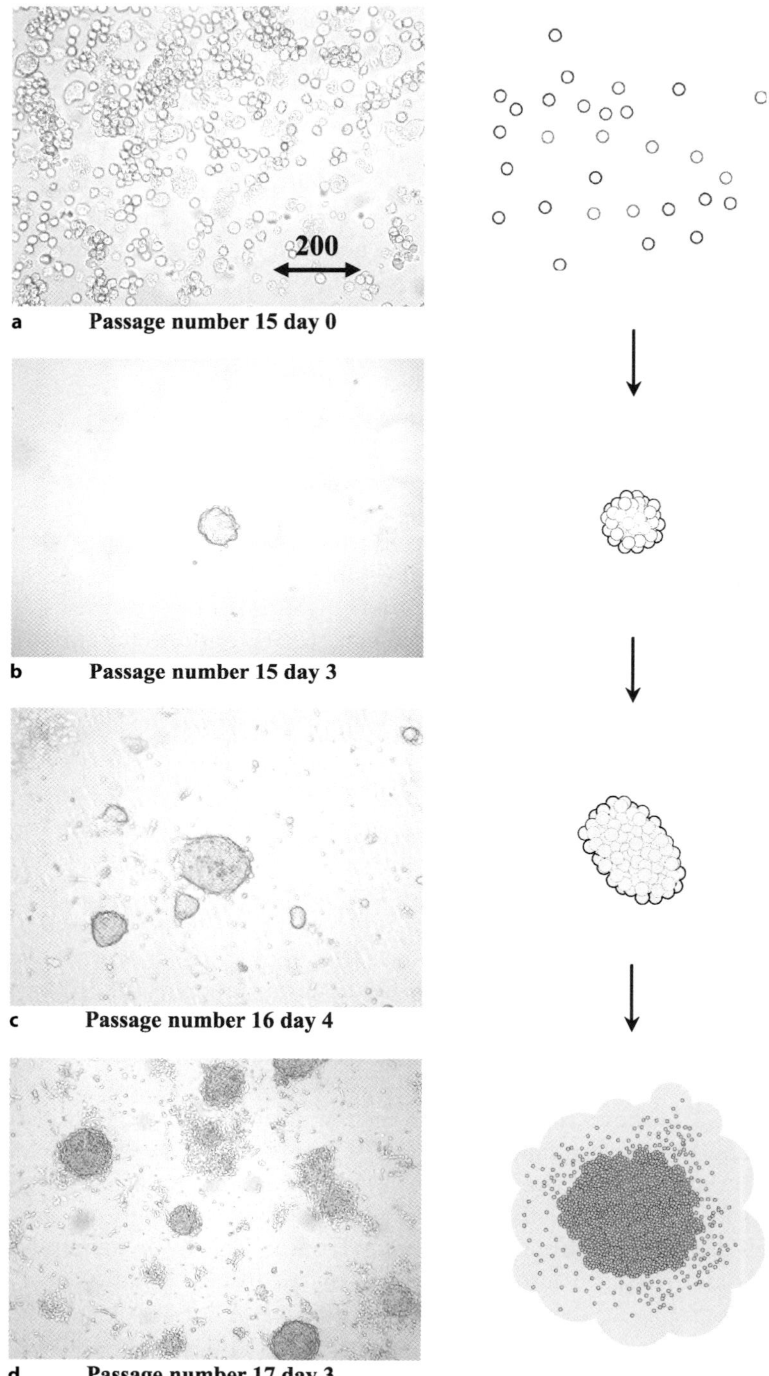

200
a   Passage number 15 day 0
b   Passage number 15 day 3
c   Passage number 16 day 4
d   Passage number 17 day 3

Fig. 5.2a–d. The optical microscopic photographs of the cultured mouse ES cells whose passage numbers were (**a**) and (**b**) 15, (**c**) 16 and (**d**) 17. The morphological change of getting flat and dark can be observed in accordance with the repeated passages and it is the typical future of the progress of the differentiation

Samples for the elemental analysis were prepared by fixing colonies that had been cultured on Mylar films with 20% formalin solution. Three and two samples are made at the passage number 16 and 17. These samples are referred to as 16-1, 16-2, 16-3, 17-1 and 17-2, respectively. The second numbers shows the period of sample preparation, and the larger number means longer period of cell culture.

5.2.2 XRF Analysis and Results

The SR-XRF analyses in this investigation were performed at the Photon Factory in beam line 4A. The incident X-ray energy was 14.3 keV and the beam size was approximately $7 \times 5 \, \mu m^2$. The detailed set-up of the beam line is described in Chap. 2. The analyses were carried out in air.

The elemental distribution images were obtained in the three or four areas that contained colonies in each sample. The typical elemental images of (b) P, (c) S, (d) Cl, (e) Fe and (f) Zn and the corresponding (a) microscopic photographs of the samples at the passage number 16 and 17 are shown in Figs. 5.3 and 5.4, respectively. The scale on the right side of the images shows the count of the X-ray intensity. Red and blue pixels show areas of high and low intensities respectively. The measurement areas were $99 \times 99 \, \mu m^2$ for Fig. 5.3 and $144 \times 144 \, \mu m^2$ for Fig. 5.4, and the measurement time was both 6 s/point. The range of intensity was from 0 to 41 for P, 0 to 44 for S, 14 to 146 for Cl, 17 to 290 for Fe and 6 to 29 for Zn in Fig. 5.3 and from 0 to 98 for P, 0 to 90 for S, 1 to 22 for Cl, 0 to 23 for Fe and 2 to 51 for Zn in Fig. 5.4, respectively. From the results of the imaging, the measurement points were selected for further quantitative point-measurement. XRF spectra were obtained at these points to reveal the distribution ratios among elements. The measurement time was 200 seconds. The typical spectra that were obtained in the colonies at the passage number 16 and 17 are shown as solid lines in Fig. 5.5a,b, respectively with the spectra obtained outside of the colonies shown as dotted lines. Each spectrum is normalized with the incident X-ray intensity. The spectra obtained in the colonies shown in Fig. 5.5a,b are compared in Fig 5.6. The solid and dotted spectra show those obtained at the passage number 16 and 17, respectively. Quantitative analysis was then applied to all measured spectra and the calculated values for concentrations of S, P, Cl, Fe and Zn are shown in Table 5.1. The thickness and the density of the colonies are considered as $50 \, \mu m$ and $1.0 \, g/cm^3$, respectively.

Table 5.1. The quantification results obtained by processing XRF spectra with the computer program that was introduced in Chap. 2. The concentrations of P, S, Cl, Fe and Zn in the mouse ES cell colonies were quantified and shown in ppm

	colony	P	S	Cl	Fe	Zn
16-1	1	1,121.2 ± 283.6	657.0 ± 188.4	651.1 ± 145.5	67.2 ± 77.4	3.4 ± 1.1
	2	1,150.2 ± 307.5	713.9 ± 201.3	318.2 ± 76.9	32.1 ± 24.3	8.7 ± 3.0
	3	1,119.7 ± 169.5	685.4 ± 90.3	461.5 ± 28.1	32.4 ± 34.6	8.2 ± 3.7
	4	1,373.4 ± 1,862.7	692.6 ± 764.0	746.5 ± 777.2	49.7 ± 55.5	2.9 ± 0.9
16-2	1	499.8 ± 161.9	411.8 ± 175.0	940.5 ± 278.0	199.3 ± 71.4	3.0 ± 0.8
	2	1,960.2 ± 461.6	1,445.8 ± 415.8	917.4 ± 204.6	981.6 ± 266.8	5.4 ± 1.3
	3	1,401.4 ± 399.9	1,129.9 ± 274.1	944.7 ± 211.3	714.5 ± 140.2	4.5 ± 1.5
	4	925.2 ± 99.6	695.3 ± 120.5	744.5 ± 54.2	426.5 ± 38.8	4.6 ± 2.1
16-3	1	1,774.1 ± 864.3	1,087.9 ± 126.2	920.3 ± 92.3	64.5 ± 61.7	12.6 ± 7.4
	2	3,421.0 ± 2,240.5	2,266.5 ± 1,277.6	1,637.4 ± 686.4	110.5 ± 61.4	37.2 ± 18.5
	3	2,666.5 ± 1,726.6	1,840.6 ± 1,027.0	1,731.9 ± 921.0	94.2 ± 47.6	40.4 ± 19.2
	4	272.2 ± 72.6	194.5 ± 57.4	247.1 ± 45.1	28.9 ± 5.5	3.0 ± 0.2
17-1	1	1,633.4 ± 825.9	1,055.9 ± 328.8	61.1 ± 5.7	52.0 ± 18.4	17.4 ± 7.1
	2	393.9 ± 170.0	277.0 ± 110.8	61.2 ± 12.1	24.6 ± 12.9	3.6 ± 1.1
	3	1,165.4 ± 967.4	772.0 ± 630.5	54.1 ± 10.6	24.7 ± 11.7	13.7 ± 10.5
17-2	1	1,409.3 ± 782.4	884.6 ± 506.2	295.1 ± 111.7	53.7 ± 33.2	8.1 ± 5.5
	2	1,986.2 ± 654.6	1,158.0 ± 316.3	476.3 ± 131.5	79.3 ± 55.5	9.3 ± 6.4
	3	2,489.2 ± 669.1	1,544.8 ± 433.1	259.6 ± 71.6	68.6 ± 106.1	23.3 ± 13.3

ES cell colony at passage number 16

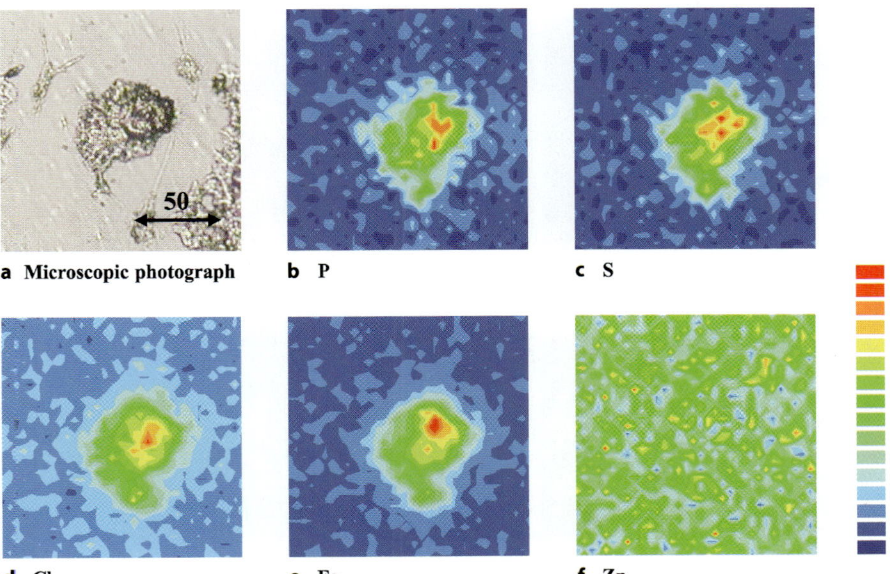

Fig. 5.3a–f. The typical elemental images of (**b**) P, (**c**) S, (**d**) Cl, (**e**) Fe and (**f**) Zn obtained in the mouse ES cell colony at passage number 16 shown in (**a**) microscopic photograph. The colony contained intracellular P, S, Cl and Fe but the concentration of Zn was low

5.2.3 Discussion

In Fig. 5.3, it can be seen that the colony whose passage number is 16 contained (b) P, (c) S, (d) Cl and (e) Fe while the distribution of (f) Zn could not be detected at a considerable level due to the low concentration. The distributions of (b) P, (c) S, (d) Cl and (e) Fe were almost identical. This result indicates that these elements are contained in this colony uniformly and suggests that the cells in this colony are in the same stage of differentiation. In Fig. 5.4, the fluorescent X-ray intensity from (f) Zn was high and that of (d) Cl was low in the colony whose passage number is 17, though it contained (b) P, (c) S and (e) Fe. The concentration of Cl had decreased and that of Zn had increased according to the progress of differentiation. And the distributions of Fe and Zn are slightly different from those of P and S. This fact suggests that the cells in this colony had differentiated into several different cell types.

From the XRF spectra shown in Fig. 5.5a,b, it was confirmed that these colonies contained the elements P, S, Cl, Fe and Zn. The peaks of Ar in these spectra were due to Ar contents in the air. Figure 5.6 shows the comparison of the typical spectra the passage number 16 (solid line) and 17 (dotted line). It can be seen that the height of the peak of Cl is low and that of Zn is high

ES cell colony at passage number 17

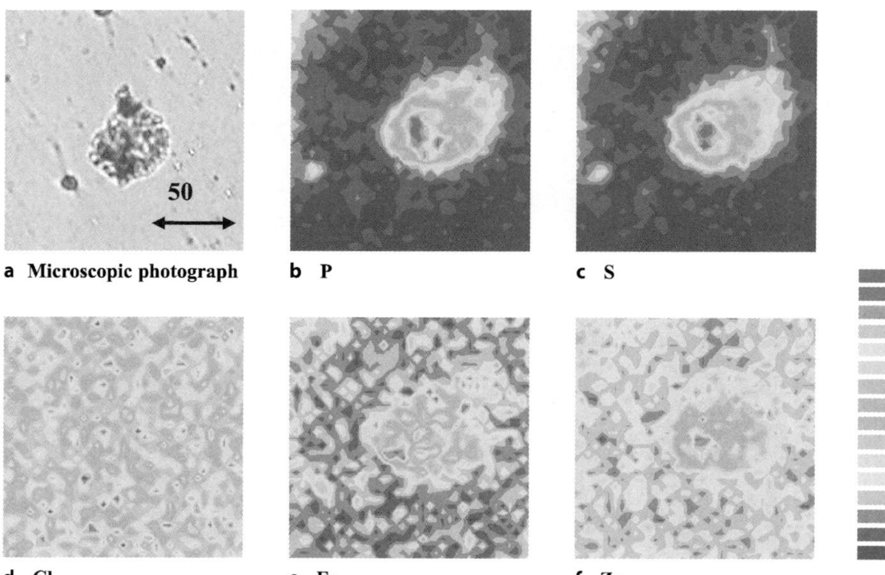

Fig. 5.4a–f. The typical elemental images of (**b**) P, (**c**) S, (**d**) Cl, (**e**) Fe and (**f**) Zn obtained in the mouse ES cell colony at passage number 17 shown in (**a**) microscopic photograph. The colony contained P, S, Fe and Zn. The distribution of Cl could not be measured due to the low concentration while Zn was low in the colony at passage number 16

in the passage number 17. This is supportive for the results of the elemental imaging.

To confirm this difference, the quantification results from each colony were compared. The absolute area densities of the intracellular elements are not so important because the sizes of the colonies were different from each other (approximately 60–200 µm in the diameter). In the present study, the relative ratios of each element to phosphorus are compared. The intracellular content of phosphorous has been considered as the index of intracellular organic content because it is a constituent of non-diffusible solutes that carries a net negative charge [20]. Figure 5.7a–c shows the relative amounts of S, Cl and Zn to phosphorus in the samples respectively. The values of S/P (Fig. 5.7a) are similar in all samples. But the values of Cl/P b had distinctly decreased and those of Zn/P c had increased in accordance with the increase of the passage number from 16 to 17.

Zn plays a key role in genetic expression, cell division, and growth in several ways and is essential for function of many enzymes [21, 22]. Mouse ES cells differentiate into a variety of cell types and the proliferation is activated as the passage is repeated. It is probable that the increase of Zn is deeply

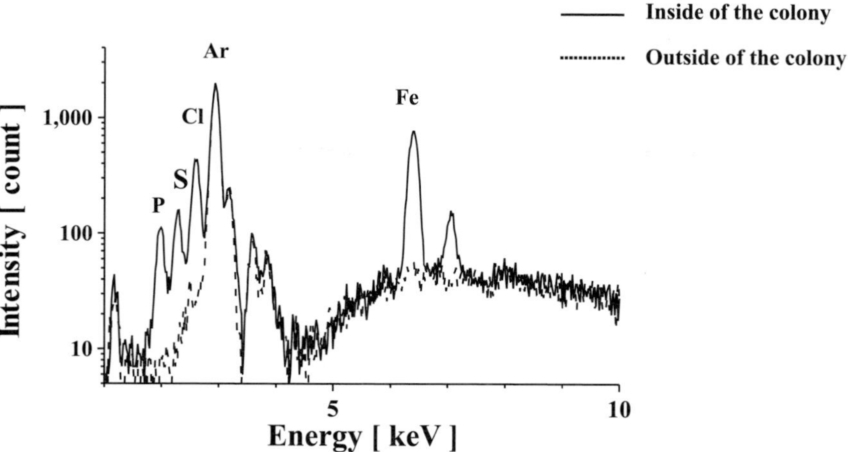

a The typical spectrum obtained in the mouse ES cell colony of the passage number 16

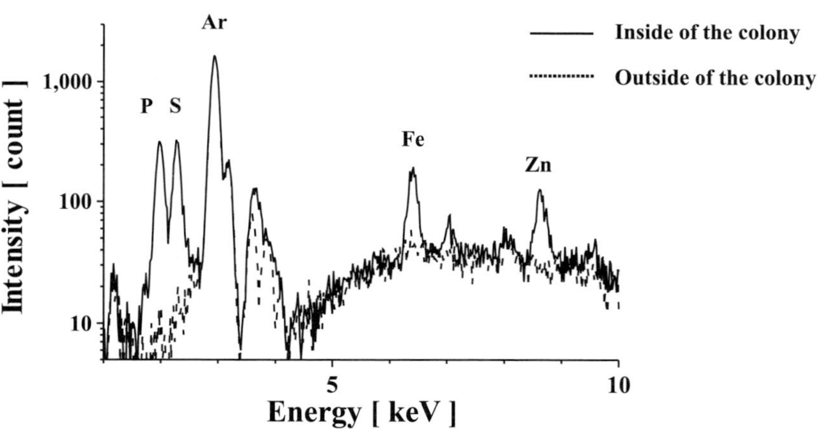

b The typical spectrum obtained in the mouse ES cell colony of the passage number 17

Fig. 5.5a,b. The typical XRF spectra obtained in the mouse ES cell colonies at the passage number (**a**) 16 and (**b**) 17. The *solid* and *dotted spectra* show those obtained inside and outside of the colonies respectively. It was confirmed that the elements such as P, S, Cl, Fe and Zn were contained in the colonies not in the culture medium

related to the growth and acquisition of cell functions. Furthermore the transcription factors with zinc-finger DNA-binding domains such as GATA-1 are required for the regulation of the differentiation into specific cells [23]. It is

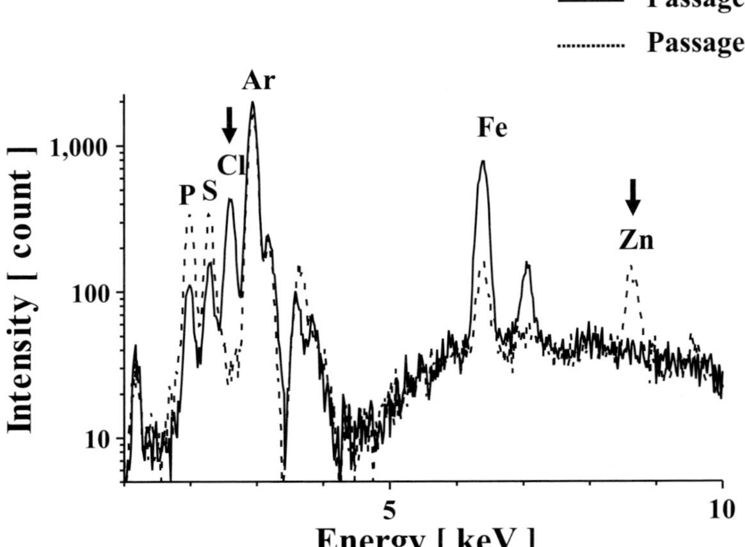

Fig. 5.6. The comparison of the typical XRF spectra obtained in the mouse ES cell colonies at the passage number 16 (*solid line*) and 17 (*dotted line*). It is suggested that chlorine had decreased and zinc had increased according to the progress of differentiation

also possible that these transcription factors had increased in mouse ES cells, as the directions of differentiation were determined in each colony.

Cl ions are mostly known to participate in the modulation of cell excitability. The Cl gradient across cell membranes adjusts the membrane potential, and it is also related to the regulation of intracellular pH and cell volume [24]. There is the possibility that these parameters had changed due to the progress of differentiation and/or the activation of cells, and the change had resulted in the decrease of Cl. In order to elucidate the effect of the change in chloride concentration, other analysis techniques should be applied complementarily for the in vivo measurement of the membrane potential or intracellular pH.

5.3 Investigation of the Process of Neuronal Differentiation

5.3.1 Induction of Neuronal Differentiation and Sample Preparation

The method suggested by Sasay et al. was adopted to induce neuronal differentiation of mouse ES cells in this study [10]. In this method, ES cells are co-cultured with the feeder layer of the mouse bone marrow-derived stromal cell line, PA6, not of the primary mouse embryonic fibroblasts [2, 10].

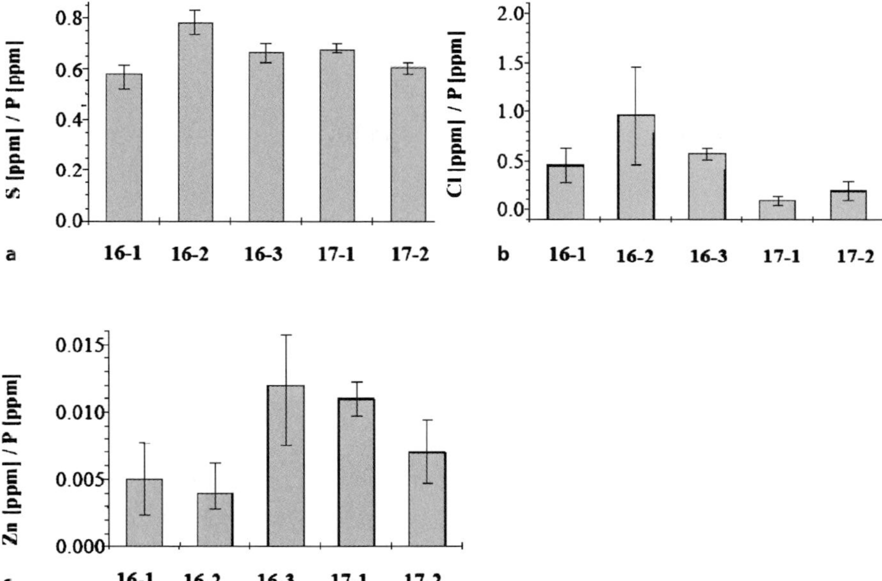

Fig. 5.7a–c. The relative amounts of (**a**) S, (**b**) Cl and (**c**) Zn to phosphorus respectively in the sample 16-1, 2, 3, 17-1 and 2. The values of S/P (**a**) are almost similar in all samples. But the values of Cl/P and Zn/P distinctly had changed in accordance with the increase of the passage number from 16 to 17

They call this activity for the generation of dopaminergic neurons "stromal cell-derived inducing activity (SDIA)". After co-culture with PA6 cells for 8 days, 92% of ES cell colonies contain differentiated neurons in their experiments. The reagents and procedures needed in this method are described in appendix 5.A. In contrast to the protocol suggested by Mckay et al., this method does not require growth in serum or the selection of neural precursor cells.

The sample was prepared by fixing the mouse ES cells (129/Sv) with 20% formalin after the co-culture with the mouse PA6 feeder layers for 8 days on the Mylar films. PA6 cells were purchased from Riken Gene Bank, the Institute of Physical and Chemical Research. Figure 5.8 shows the optical microscopic photograph of the ES cell colony that had been co-cultured with PA6 for 8 days. The dendrite-like tissue that had widely spread out of the colony can be seen and is the morphological feature of the neurons.

The sample referred as 16-3 in the previous paragraph was also investigated with XANES technique as the undifferentiated control sample.

5.3.2 Experimental Procedures and Results

The SR-XRF elemental imaging and XANES analysis in this investigation were performed at the beam line 39XU of SPring-8. The incident X-ray energy

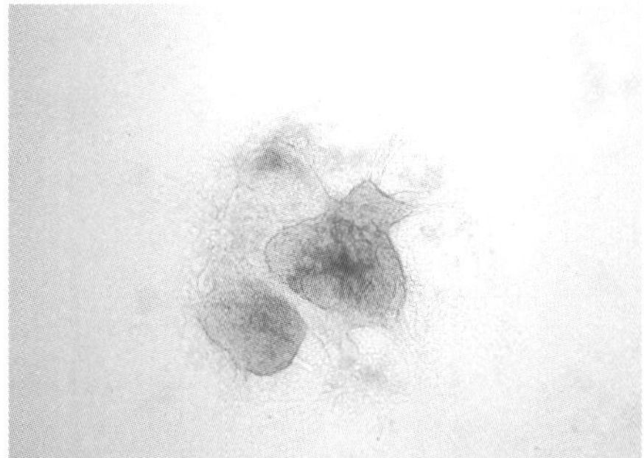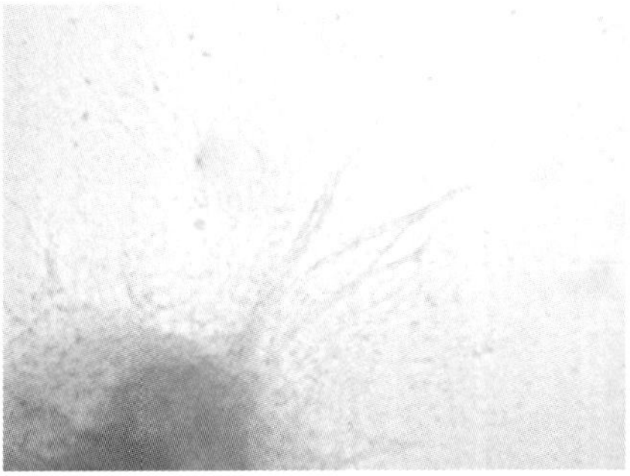

Fig. 5.8. The optical microscopic photograph of the cultured mouse ES cells after co-culture with the feeder layer of the bone marrow-derived stromal cell line, PA6. The dendrite-like tissue that had widely spread out of the colony can be seen and is the morphological feature of the neurons

was $8.7\,\text{keV}$ in XRF analysis. The detailed set-up of the beam line is described in Chap. 2. The analyses were carried out in vacuum.

The elemental distribution images of the colony of the sample were obtained. The elemental images of (a) P, (b) S, (c) Cl, (d) Fe, (f) Zn and the corresponding microscopic photograph of the sample (a) are shown in Fig. 5.9. The scale on the right side of the images shows the count of the X-ray intensity. The measurement area was $280 \times 320\,\mu\text{m}^2$ and the measurement time was $3\,\text{s/point}$. The range of intensity was from 154 to 932 for P, 6 to 291 for S, 4 to 94 for Cl, 10 to 211 for Fe and 3 to 130 for Zn, respectively. The beam size was $10\,\mu\text{m}$ in the diameter.

From the results of the imaging, the points with the highest fluorescent intensity of Zn and Fe were selected for XANES analysis. The Zn and Fe K-edge XANES spectra obtained at the measurement points are shown in Fig. 5.10a,b, respectively. These spectra were measured in fluorescence mode. The Zn K-edge absorption spectrum was collected from 9.72 to $9.63\,\text{keV}$ with an energy resolution of $0.5\,\text{eV}$ and energy shifts of resolved absorption peaks of $0.25\,\text{eV}$ was detected. The spectrum was collected with a total signal averaging of $60\,\text{s/point}$. The Fe K-edge absorption spectrum was collected from 7.16 to $7.10\,\text{keV}$ with an energy resolution of $0.5\,\text{eV}$ and the measurement time was $70\,\text{s/point}$. Each spectrum represents the ratio I_f/I_0 (I_f = fluorescence counts, I_0 = photon incident flux measured by ionization chambers) as a function of photon energy. The XANES spectra from the colony in the control sample are shown in these figures a the dotted lines. The measurement time was $30\,\text{s/point}$ and $90\,\text{s/point}$ for the Zn and Fe K-edge absorption spectra respectively. The spectra obtained from reference samples (Zn, ZnO, FeO, Fe_2O_3 and Fe_3O_4) are also shown as thin color lines. The spectra of reference samples were collected in transmission mode and each spectrum represents the value of $-\exp(I/I_0)$ (I = transparence counts, I_0 = photon incident flux) as function of photon energy. The measurement times for these

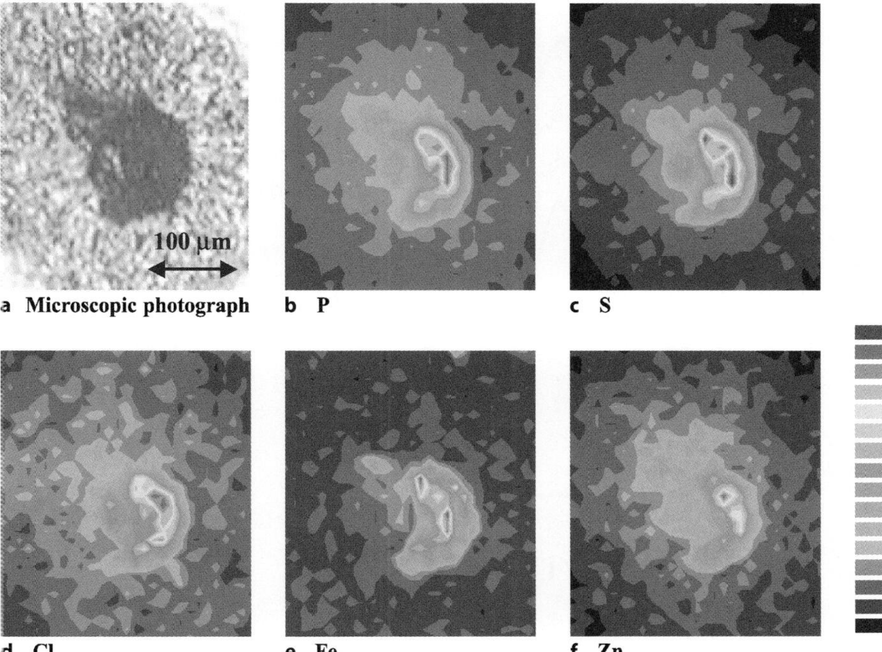

Fig. 5.9a–f. The elemental images of (**b**) P, (**c**) S, (**d**) Cl, (**e**) Fe and (**f**) Zn obtained in the mouse ES cell colony after co-culture with PA6 for 8 days that is shown in (**a**) microscopic photograph. The colony contained intracellular P, S, Cl, Fe and Zn

reference samples were 3 s/point. All spectra are normalized with respect to both maximum and minimum intensities.

XRF spectrum of the sample co-cultured with PA6 was also obtained at Photon Factory in beam line 4A and is shown as Fig. 5.11. The dotted line shows the spectrum obtained outside of the ES cell colony but at the PA6 feeder layer. The incident X-ray energy was 7.2 keV, and the beam size was approximately $500 \times 500\,\mu m^2$. The analysis was carried out in vacuum. The spectra were collected with a total signal averaging of 200 seconds. Quantitative analysis was then applied to the spectrum obtained in ES cell colony and the calculated concentrations of S, P, Cl and Fe are shown in Table 5.2. The volume of the colony was regarded as the ellipsoid ($d_x, d_y = 100, d_z = 50\mu m$) and the density was considered as $1.0\,g/cm^3$.

5.3.3 Discussion

Figure 5.9 shows the colony that was co-cultured with PA6 contained (b) P, (c) S, (d) Cl, (e) Fe and (f) Zn. The distribution of Fe and Zn has a minor difference from those of other nonmetal elements while the distributions of P,

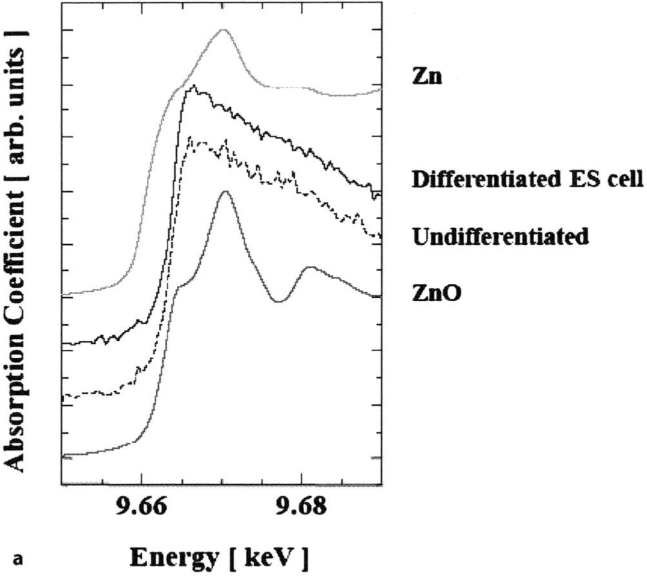

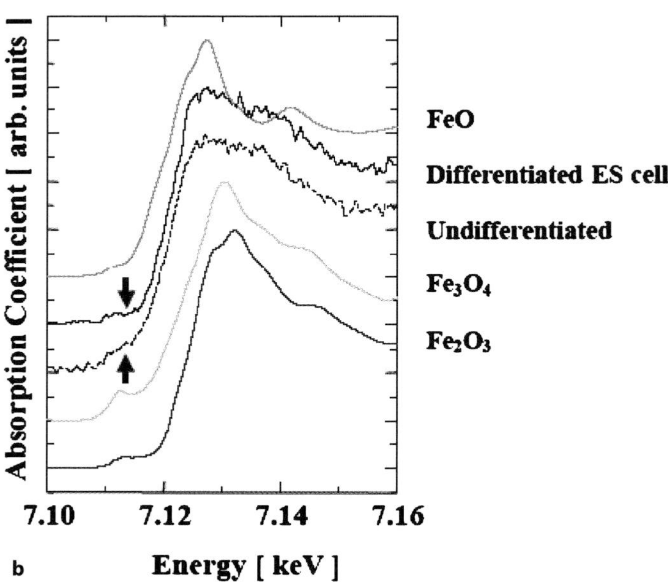

S and Cl are identical. This variety of distributions suggests the heterogeneity of intracellular constituents and that the cells in this colony had differentiated into several different kinds of cells. Sasai et al. indicates the possibility of the production of less than 2% glial or mesodermal lineages, and dopaminergic,

Fig. 5.10a,b. The (**a**) Zn and (**b**) Fe K-edge XANES spectra obtained at the mouse ES cell colony after co-culture with PA6 for 8 days (*solid line*) and the colony in the undifferentiated control (*dotted line*). The spectra collected from reference samples (Zn, ZnO, FeO, Fe_2O_3 and Fe_3O_4) are also shown. No distinct difference can be seen between the solid and dotted spectra in (**a**), but at the rising edge region in (**b**), the dotted spectrum is slightly higher than the solid spectrum. It can be seen that these colonies contained the elements such as P, S, Cl, Fe and Zn

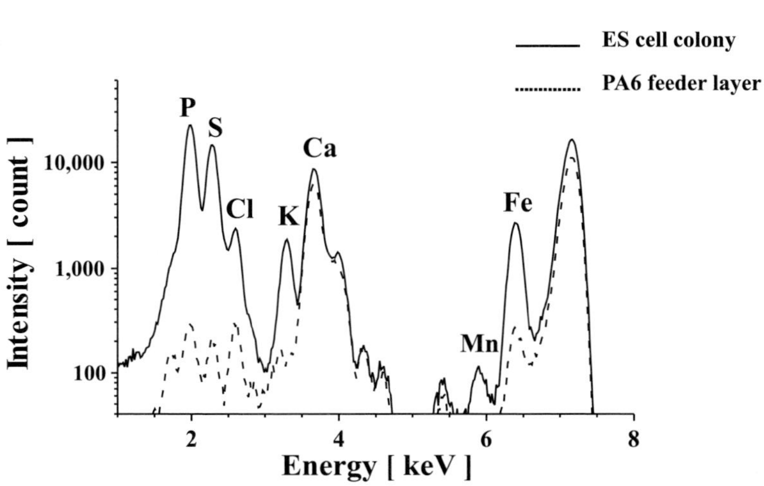

Fig. 5.11. The XRF spectra obtained in the mouse ES cell colonies after the co-culture with PA6 for 8 days (*solid line*) and at the PA6 feeder layer (*dotted line*). It can be seen that these colonies contained the elements such as P, S, Cl, Fe and Zn

Table 5.2. The quantification results obtained by processing XRF spectrum of the mouse ES cell colony after the co-culture with PA6 for 8 days. The concentrations of P, S, Cl, K, Mn and Fe were quantified and shown in ppm

P	S	Cl	K	Mn	Fe
16,041.0	6,599.1	637.0	300.3	4.1	93.3

cholinergic, and serotonergic neurons in their experiments [10]. It can be considered that the differences among neuronal functions are related to the concentrations of intracellular elements.

In Zn K-edge XANES spectra shown in Fig. 5.10a, no significant difference was seen between the spectrum obtained at the colony co-cultured with PA6 cells (solid line) and that from the undifferentiated control (dotted line). This result indicates that the local structure such as inter-atomic distances and coordination number or Zn contents is similar in these samples. From the comparison of the spectra with those of reference samples,

Zn was in the form of Zn^{2+} as ions and/or metalloproteins in these samples. In Fe K-edge XANES spectra shown in Fig. 5.10b; however, the spectrum obtained in the colony of undifferentiated control (dotted line) shows higher values than that from the colony co-cultured with PA6 cells (solid line) at the rising edge region (7.110–7.115 keV). The capacity of readily exchanging electrons makes iron essential for fundamental cell functions and metalloproteins containing iron play an important role in controlling protein synthesis and cell differentiation [25, 26]. The difference in iron stereochemistry is considered to represent the difference in the functions of the proteins bound to iron. It is also possible that the free iron was contained in these ES cell colonies in different states. In both cases, the comparison of the absorption edge with reference samples shows that iron is contained as Fe^{2+} in these samples.

XRF spectrum in Fig. 5.11 indicates that the differentiated ES cell colony (solid line) contained P, S, Cl and Fe as well as the undifferentiated colony that was investigated in the previous experiments (Fig. 5.5a,b). After the co-culture with PA6 cells, however, the fluorescent X-ray peaks from K and Mn can be seen in the spectrum obviously. It is considered that these elements had been imported into the cell during the culture on the PA6 feeder layer. K and Mn are characteristic and essential constituents in neurons. Mn is essential for protein synthesis, normal development and activity of the nervous tissue and prevents apoptosis induced by Fe^{2+}, amyloid beta-peptide and nitric oxide-generating agents as the active center of Mn superoxide dismutase (MnSOD) [27, 28]. K ions are widely utilized to control electric activities of neurons and are deeply related to the regulation of action potential, membrane potential and cell volume in every kind of cells. It was revealed that the acquisition of these elements had been occurred in accordance with the neuronal differentiation of ES cells.

The XRF spectrum obtained from PA6 feeder layer shows that the PA6 cells contained K in addition to P, S, Cl and Fe. It is considered that the ES cells had received K from the feeder layers or the feeder layer had played the role to promote the absorption of K by ES cells. It is clear that ES cells had not spontaneously absorbed these elements from the culture medium because the constituents of the medium were almost the same as those in the experiments about unoriented differentiation and they did not contain K and Mn without PA6 cells. This result indicates the possibility that SDIA plays the role to promote the uptake of K and Mn.

The relative ratios of S, Cl to P (S/P are Cl/P) are calculated from the quantification results (Table 5.2) as 0.41 and 0.04 respectively. These values are much lower than those of ES cell colonies cultured without PA6 feeder layers. The decrease of Cl is possible to be related to the loss of pluripotency during the progress of differentiation as seen in the experiments about the effect of unoriented differentiation.

5.4 Conclusion

In this study, the role of intracellular trace elements in the differentiation of mouse ES cells was investigated by analyzing the distribution, concentration and chemical states in the process of differentiation using micro-SR-XRF and XANES analysis. The elemental information at the cell level is important to elucidate the changes in ES cells during the differentiation and reveal the optimal elemental conditions for neuronal differentiation.

As the first experiment, the change of elemental distributions and concentrations in mouse ES cell colonies during the unoriented differentiation were analyzed. The mouse ES cell colonies whose passage numbers were 16 and 17 were investigated. The results of XRF elemental imaging revealed that the distributions of each element become heterogeneous according to the increase of passage number. This heterogeneity suggests that cells in the passage number 17 had differentiated into a variety of cell types in the colonies. The imaging analysis also suggested the increase of Zn and the decrease of Cl according to the passage. This tendency was confirmed by quantitative XRF point analysis. The concentrations of Zn had increased from 0.07 to 0.09 and those of Cl had decreased from 0.67 to 0.14 in the average of relative amounts to P during the change of the passage number. The difference among concentrations of these elements is considered to be deeply related to the biological functions such as proliferation, gene transcription and cell excitability in the process of the differentiation.

In the second experiment, the mouse ES cells that had been co-cultured with PA6 feeder layer for 8 days were analyzed with XRF and XANES technique. XRF analysis showed that the elemental distributions of Fe and Zn were different from those of P, S and Cl. This result indicates the possibility of the production of several different cell types such as dopaminergic, cholinergic, and serotonergic neurons. XANES spectrometry revealed that the chemical state of Zn in the colony after co-culture with PA6 was similar to that in undifferentiated control. On the other hand, the local structure around Fe had been slightly changed due to the neuronal differentiation induced by PA6 feeders. Furthermore it was revealed by XRF spectrum that the colony contained K and Mn in addition to P, S, Cl and Fe. These are characteristic and essential components for neurons. It is considered ES cells had absorbed these elements during the neuronal differentiation promoted by the PA6 feeder layer. Quantitative analysis applied to this spectrum showed the decrease of S and Cl in the relative concentrations to P.

Present study demonstrated that the intracellular elemental conditions such as distribution, concentration and chemical states reflect the specific biological functions. The distinct differences could be observed between the constituents of mouse ES cells with and without the induction of neuronal differentiation.

The identification of cell types into which the mouse ES cells were differentiated was not carried out in this experiment. The information about the result of differentiation can be obtained by the histochemical approach with the antibody staining after the elemental analysis. The complemental analysis such as histological observation is possible due to the non-destructivity of XRF and XANES analysis. By identifying the cell types histochemically that correspond to the elemental conditions and structuring the database, it would be possible to know the cell types that are contained in embryonic body and the effect of the differentiation by the elemental analysis.

It was revealed that elements such as Cl, K, Mn, Fe and Zn behave in a unique manner and are considered to play an important role during the differentiation. This finding indicates the possibility that the differentiation of ES cells into the specific cell type can be regulated by stimulating cells with the reagents that affect the concentrations and chemical states of these intracellular elements. It is well recognized that deficiencies or excesses of essential trace elements during early development give mouse ES cells the significant effects such as structural abnormalities or embryonic death through the direct metal binding to critical membrane sites or intracellular ligands including protein and nucleic acids [15]. It is considered that modified quantity of adequate reagents has the positive and selective effect on ES cells.

The present study also confirmed that XRF spectrometry is the powerful method to investigate the multiple intracellular trace elements efficiently at the high sensitivity and resolution. XRF and XANES analysis can be applied in combination with other techniques such as staining or patch-clamp recording. These X-ray analysis methods should be utilized in the further investigation that is needed to explain the mechanism of the differentiation of embryonic stem cell.

References

1. S. Gokhan, M.F. Mehler, *Anat. Rec.,* **2001**, 265, 142.
2. M. Hynes, A. Rosenthal, *Neuron,* **2000**, 28, 11.
3. S.B. Dunnett, A. Bjorklund, O. Lindvall, *Nature Rev. Neurosci.,* **2001**, 2, 365.
4. M.G. Klug, M.H. Soonpaa, G.Y. Koh, L.J. Field, *J. Clin. Invest.,* **1996**, 98, 216.
5. M. Kennedy, M. Firpo, K. Chol, C. Wall, S. Robertson, N. Kabrun, G. Keller, *Nature,* **1997**, 386, 488.
6. G. Keller, M. Kennedy, T. Papayannopoulou, M.V. Wiles, *Mol. Cell. Biol.,* **1993**, 13, 473.
7. J.A. Robertson, *Nature Rev. Genet.,* **2001**, 2, 74.
8. A. Fraichard, O. Chassande, G. Bilbaut, C. Dehay, P. Savatier, J. Samarut, *J. Cell Sci.,* **1995**, 108, 3181.
9. S.H. Lee, N. Lumelsky, L. Studer, J.M. Auerbach, R.D. McKay, *Nature Biotech.,* **2000**, 18, 675.
10. H. Kawasaki, K. Mizuseki, S. Nishikawa, S. Kaneko, Y. Kuwana, S. Nakanishi, S. Nishikawa, Y. Sasai, *Neuron,* **2000**, 28, 31.
11. M. Lako, S. Lindsay, J. Lincoln, P.M. Cairns, L. Armstrong, N. Hole, *Mech. Dev.,* **2001**, 103, 49.

12. J.F. Loring, J.G. Porter, J. Seilhamer, M.R. Kaser, R. Wesselschmidt, *Restor. Neurol. Neurosci.,* **2001**, 18, 81.
13. A. Streit, A.J. Berliner, C. Papanayotou, A. Sirulnik, C.D. Stern, *Nature,* **2000**, 406, 74.
14. V. Tropepe, S. Hitoshi, C. Sirard, T.W. Mak, J. Rossant, D. Van der Kooy, *Neuron,* **2001**, 30, 65.
15. L.A. Hanna, J.M. Peters, L.M. Wiley, M.S. Clegg, C.L. Keen, *Toxicol.,* **1997**, 116, 123.
16. K.S. O'Shea, *Anat. Rec.,* **1999**, 257, 32.
17. M.J. Evans, M.H. Kaufman, *Nature,* **1981**, 292, 154.
18. G.R. Martin, *Proc. Natl. Acad. Sci. USA,* **1981**, 78, 7634.
19. *"Animal Cell Electroporation and Electrofusion Protocols"*, *Methods in molecular biology*, vol. 48, ed. J.A. Nickoloff, Humana Press, **1995**, 167.
20. S. Larsson, A. Aperia, C. Lechene, *Am. J. Physiol.,* **1986**, 251, C455.
21. H.H. Sandstead, *J. Lab. Clin. Med.,* **1994**, 124, 322.
22. R.S. MacDonald, *J. Nutr.,* **2000**, 130, 1500S.
23. L. Pevny, M.C. Simon, E. Robertson, W.H. Klein, S.F. Tsai, V.D. Agati, S.H. Orkin, F. Constantini, *Nature,* **1991**, 349, 257.
24. L. Garcia, M. Rigoulet, D. Georgescauld, B. Dufy, P. Sartor, *FEBS Lett.,* **1997**, 400, 113.
25. G. Cairo, A. Pietrangelo, *Biochem. J.,* **2000**, 352, 241.
26. P. Ponka, *Am. J. Med. Sci.,* **1999**, 318, 241.
27. J.N. Keller, M.S. Kindy, F.W. Holtsberg, D.K. St Clair, H.C. Yen, A. Germeyer, S.M. Steiner, A.J. Bruce-Keller, J.B. Hutchins, M.P. Mattson, *J. Neurosci.,* **1998**, 18, 687.
28. M.P. Mattson, Y. Goodman, H. Luo, W.M. Fu, K. Furukawa, *J. Neurosci. Res.,* **1997**, 49, 681.

5.A Appendix: Culture Procedure of Mouse ES Cell

In this study, the cell culture procedure recommended by Mudgett and Livelli [1] was modified according to other references [2–4].

5.A.1 Culture of Mouse ES Cell without Feeder Layers of PMEF

Requirements

Mouse ES cell (129/Sv) passage number 15, 2.5×10^6 cells/vial, 1 vial
  Mouse ES cell medium
  Knockout DMEM 200 ml
  Knockout serum replacement (KSR) 37 ml
  L-glutamine 2.5 ml
  Non-essential amino acid (NEAA) 2.5 ml
  Penicillin-Streptomycin 2.5 ml
  2-Mercaptoethanol (2-ME) 2.5 ml
  Leukemia inhibitor factor (LIF): ESGRO, 10^6 unit/ml, 250 µl

0.1% gelatin
Trypsin/EDTA
Phosphate buffer saline (PBS)
Dimethyl sulfoxide (DMSO)

KSR and Knockout DMEM were purchased from Invitrogen, Inc., ESGRO was purchased from CHEMICON International, Inc., all other reagents and mouse ES cells were purchased from Cell & Molecular Technologies, Inc. via Cosmo Bio, Inc.

Knockout serum replacement (KSR), a defined serum-free supplement, has been developed to maintain ES cells in undifferentiated state because the main source of potential differentiating factors in ES cell culture is fetal bovine serum (FBS) [2]. Knockout DMEM is an optimized DMEM formulation for the use of KSR. Purified recombinant leukemia inhibitor factor (LIF) prevents the differentiation of ES cells in a dose-dependent manner [5]. To maintain the pluripotency of the mouse ES cells, it is recommended to use recombinant LIF in combination with a primary mouse embryonic fibroblast (PMEF) feeder layer. PMEF layers produce unspecified factors and aid in keeping the ES cells from differentiating.

Thaw mouse ES cell

1. Treat 6 mm tissue culture dishes with sterile 0.1% gelatin. Swirl 3 ml gelatin to fully cover the dish and incubate it for at least 30 min. Aspirate the gelatin solution and discard. There is no need to dry the dishes following treatment.
2. Thaw a vial ES cells in 37 °C water bath into the mixture of 4 ml ES media and 4 ml KSR that have been warmed to 37 °C. Centrifuge at 1000 rpm for 5 min, aspirate the media and resuspend the cells in 10 ml ES media. Plate the ES cells onto 2 gelatinized dishes. As DMSO used as the freezing media is toxic for cells, thawing should be finished quickly.
3. Check the cells the next day to determine if fresh media is required. The change of media color to yellow indicates deterioration of media.
4. Check again the next day, replace with fresh media if required. Once the dish is crowded or colonies are large, passage 1 in 2 dishes. It takes 2 or 3 days from thawing. If the dish is too crowded, cells tend to differentiate.
5. The addition of media is required every 1–2 days.

Passage mouse ES cell

1. Prepare gelatinized dishes. After removing ES media and rinsing dishes with PBS twice, swirl 1 ml Trypsin/EDTA and discard the solution. This process is to remove the serum or KSR that blocks the effect of Trypsin. Add another 1 ml Trypsin/EDTA and discard the solution again. Then incubate at 37 °C for 3–5 min. Attend not to dry the surface of the cells.

2. Add 10 ml ES media and pipette vigorously to break up all the clumps. ES cells tend to stick together, which makes it difficult to count.
3. Add 5 ml to each of two gelatinized dishes. If there are abundant dishes with ES cells, freeze and keep them in liquid nitrogen.

Freeze mouse ES cell

1. Change the media 3 hours before freezing and keep the cells active.
2. Remove ES media, wash dishes with PBS twice, and add and swirl 1 ml Trypsin/EDTA and discard the solution. Add another 1 ml Trypsin/ EDTA and discard the solution again. Then incubate at 37 °C for 3–5 min. This process is the same as the passage.
3. Add 10 ml ES media and pipette vigorously to break up all the clumps and centrifuge at 1000 rpm for 5 min.
4. Aspirate the media and resuspend cells in the appropriate volume of freezing media, the mixture of 90% ES cell medium +10% DMSO. Count cells on a hemocytometer and pipette the ES cells into cryovials at between 2×10^6 and 1×10^7 cells/vial.
5. Close caps of the cryovials firmly and cover them with foamed polystyrene and plastic bags to prevent rapid change in temperature. Figure 5.12 shows the schematic drawing of this procedure.
6. Freeze them immediately in the deep freezer (-80 °C) and transfer them into liquid nitrogen the next day. If the deep freezer is not available, place them into the freezer (-20 °C) for 2 hours and then transfer them into liquid nitrogen as covered. Freezing of -1 °C/min is desirable to optimize the cell rise.
7. It is important to label the cells with the correct passage number.

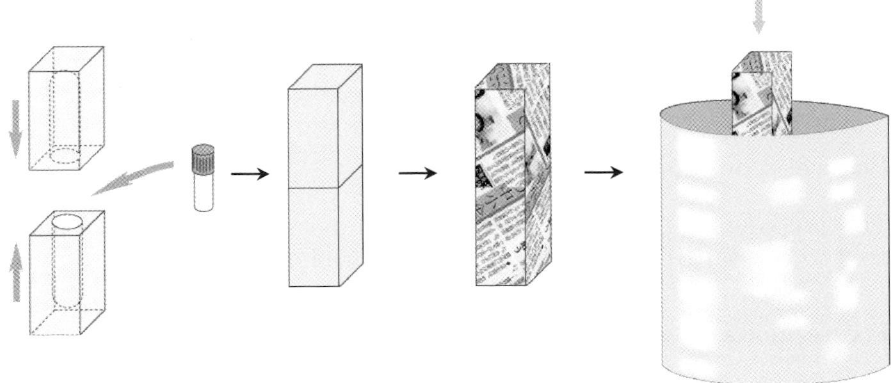

Fig. 5.12. The schematic drawing of the pretreatment in freezing ES cells that have been pipetted into the cryovials. Close caps of the cryovials firmly and cover them with foamed polystyrene, papers and plastic bags to prevent rapid change in temperature. Freezing of -1 °C/min is desirable to optimize the cell rise

5.A.2 Preparation of Mouse PA6 Cell Feeder Layer

Requirements

Mouse MC3T3-G2/PA6, 1 vial
   PA6 medium
   MEMα 90 ml
   Fetal bovine serum (FBS) 10 ml
   Trypsin/EDTA
   Dimethyl sulfoxide (DMSO)
   Mouse MC3T3-G2/PA6 was purchased from RIKEN Gene Bank, the Institute of Physical and Chemical Research. MEMα and FBS were purchased from Tech-Line, Inc. All other reagents and mouse ES cells were purchased from Cell & Molecular Technologies, Inc. via Cosmo Bio, Inc.

Thaw mouse MC3T3-G2/PA6 cell

1. Treat 6 mm tissue culture dishes with sterile 0.1% gelatin. Swirl 3 ml gelatin to fully cover the dish and incubate it for at least 30 min. Aspirate the gelatin solution and discard. There is no need to dry the dishes following treatment.
2. Thaw a vial MC3T3-G2/PA6 cells in 37 °C water bath into 10 ml PA6 media that have been warmed to 37 °C. Centrifuge at 1000 rpm for 5 min, aspirate the media and resuspend the cells in 3 ml ES media. Plate the ES cells onto 3 gelatinized dishes. Add 4 ml PA6 media to each dish (5 ml in total).
3. Check the cells the next day to determine if fresh media is required.
4. According to the maturation of PA6 cells, they adhere to the bottom of dish. Once the bottom is crowded, it can be used as the feeder layer for the culture of mouse ES cell. It takes 3 or 4 days from thawing. If it is not used as the feeder layer, passage 1 in 4–5 dishes.
5. The addition of media is required every 2–3 days.

Passage mouse MC3T3-G2/PA6 cell

This procedure is the same as the passage of mouse ES cells. Dishes don't need to be gelatin-coated. Use PA6 media instead of ES cell media.

Freeze mouse MC3T3-G2/PA6 cell

This procedure is the same as the freezing of mouse ES cells. Use PA6 media instead of ES cell media. Freezing media is the mixture of 90% PA6 medium +10% DMSO.

5.A.3 Neural Differentiation from Embryonic Stem Cells

Requirements

Mouse ES cell
   Mouse PA6 feeder layer
   Differentiation medium
   Knockout DMEM 150 ml
   L-glutamine 1.875 ml
   Non-essential amino acid (NEAA) 1.875 ml
   2-Mercaptoethanol (2-ME) 1.875 ml
   Knockout serum replacement (KSR) 17.3 ml

Induction of neuronal differentiation

1. Aspirate the PA6 media from the dish in which PA6 feeder layer is cultured. Rinse twice with the differentiation media to remove FBS from medium since the addition of FBS and LIF strongly inhibit neural differentiation.
2. <Passage ES cells on the PA6 feeder layer> Prepare mouse ES cells. Remove ES media, wash dishes with PBS twice, and add and swirl 1 ml Trypsin/EDTA and discard the solution. Add another 1 ml Trypsin/ EDTA and discard the solution again. Then incubate at 37 °C for 3–5 min.
3. Add 10 ml differentiation media and pipette vigorously to break up all the clumps.
4. Add 5 ml to each of two dishes with PA6 feeder layers. Medium change is performed on next day and every other day following that.
5. ES cells are cultured on PA6 for 8 days in differentiation medium to induce the neuronal differentiation. It is important not to add too many ES cells because they proliferate exponentially during the co-culture for 8 days.

5.A.4 Culture on Mylar Film and Sample Preparation

Sterilization of Mylar films

1. Cut the Mylar film into sections of 4×4 cm^2 square and wipe with ethanol for disinfection thoroughly.
2. Fill the dish with 70% ethanol and soak film sections for 12 hours. 70% ethanol should be prepared with pure ethanol and ultra-pure water.
3. Fill the dish with 80% ethanol and soak them for 12 hours.
4. Fill the dish with 90% ethanol and soak them for 6 hours.
5. Fill the dish with pure ethanol and soak sections, uncover the dish and dry them under ultraviolet irradiation in the clean bench.
6. After rinsing the Mylar section with ultra-pure water, place on a new dish then use for the cell culture.

Fixation with 20% formalin

1. Aspirate the media from the dish and discard. Then rinse the cells on the Mylar film with 4 °C PBS and ultra-pure water twice for each. Attend not to subject cells directly to infusions and pipette them on the side of the dish because some cells are easy to be detached.
2. Fill the dish with 4 °C 20% formalin and place it in a refrigerator (4 °C) for 30 min. Swirl the dish every 10 min.
3. Aspirate the solution and discard into the toxicant storage. After rinsing with ultra-pure water, dry them in the clean bench.
4. Cut the Mylar film into appropriate sizes and embed on the acrylic board with Kapton tape.

Appendix References

1. *"Methods in Molecular Biology, Vol.48: Animal Cell Electroporation and Electrofusion Protocols"*, ed. J.A. Nickoloff, Humana Press Enc.
2. M.D. Goldsborough, M.L. Tilkins, P.J. Price, J.L. Alfonso, J. R. Morrrison, M.E. Stevens, J. Meneses, R. Pedersen, B. Koller, A. Latour, *Focus*, **1998**, 20, 8.
3. Website of CHEMICON International, Inc., http://www.esgro-lif.com/esgro/protocols.htm
4. H. Kawasaki, K. Mizuseki, S. Nishikawa, S. Kaneko, Y. Kuwana, S. Nakanishi, S. Nishikawa, Y. Sasai, *Neuron*, **2000**, 28, 31.
5. S. Pease, P. Braghetta, D. Gearing, D. Grail, R.L. Williams, *Dev. Biol.*, **1990**, 141, 344.

6

Investigation
of Neurodegenerative Disorders (I)

6.1 Introduction

Metallic elements, especially transition elements, appear to play an important role in neurodegenerative disorders such as Parkinson's disease (PD), Alzheimer's disease (AD), amyotrophic lateral sclerosis (ALS) and Guamanian parkinsonism-dementia complex (PDC) [1–5], based on findings of excessive accumulations of metallic elements in the brains of PDC cases, and iron in idiopathic Parkinson's disease [6–9] and Alzheimer's disease [10,11]. The pathogenesis of the disease is still unknown, and researchers have suggested environmental factors as being of etiological importance [12,13]. The "oxidative stress" hypothesis suggests that excessive transition metals can exchange electrons and change their valence, which would promote production of free radicals that cause oxidative damage and neuronal degeneration [14–18]. Oxidative stress is defined as the strain of cellular function induced by reactive oxygen species (ROS), such as superoxide anions (O_2^-), hydroxyl radicals (OH^-), hydrogen peroxide (H_2O_2), and peroxynitrite ($ONOO^-$) [19]. ROS can induce lethal cellular damage through oxidation and peroxidation of proteins, lipids, and nucleic acids. Details of the relationships between transition metals and neuronal degeneration are still unclear. The distribution and chemical state of these metals is expected to give some insights into the understanding of pathogenic mechanism of these neurodegenerative disorders.

The distribution and metabolism of metallic elements such as iron have been investigated by using spectroscopic methods such as energy dispersive X-ray electron microprobe analysis [8,20], laser microprobe mass analyzer (LAMMA) [9], and nuclear microprobe [21]. These studies demonstrated the localization and distribution of iron in tissues from a patient with neurodegenerative disease, but these techniques cannot analyze the chemical state of the iron without homogenization or isolation.

Synchrotron radiation X-ray fluorescence (SR-XRF) spectrometry provides an alternative, powerful new tool for the investigation of trace elements [22]. It enables non-destructive analysis of the distributions, concentrations and chemical states at the cellular level of biological specimens with special high sensitivity and resolution [23–25]. Beside the fact that no fragmentation of the samples is required, the main advantage of this method lies in its non-invasive nature, as X-ray beams do not damage the samples. Therefore it is possible to conduct histological or histochemical analyses by staining after the elemental analysis [26–28].

XANES spectroscopy provides the means to obtain information on the valence state and binding structure of the absorbing elements, even at trace levels. XANES spectroscopy has been used for chemical state analyses on a wide range of biological samples [29,30] and pathological specimens [31,32].

Synchrotron radiation methods, thus, enable a new approach in studies about the mechanism of neurodegenerative disorder and cell death by combining the quantitative information of trace metal elements and histological observation at the single cell level. The information obtained in this analysis is significant not only from the viewpoint of therapeutic and medical applications, but also the basic biology concerning cell death and function of proteins. In this and the following chapter, investigations of parkinsonism-dementia complex, amyotrophic lateral sclerosis and Alzheimer's disease are presented as examples of applications of SR in neurodegenerative disorders. The pathology of these diseases is extremely complex and the decisive mechanism responsible for the neuronal cell death remains to be elucidated completely. However, SR-based studies are expected to provide innovative data and knowledge about these disorders through high sensitivity elemental and chemical state analysis.

6.2 Parkinsonism-Dementia Complex

6.2.1 Introduction

Parkinsonism-dementia complex (PDC) and amyotrophic lateral sclerosis (ALS) are prominently found at the three foci in the western Pacific region (Guam, Kii Peninsula and West New Guinea). The neuropathology is characterized by severe neuronal loss and by widespread neurofibrillary tangles (NFT) in affected patients [12]. A major consideration in assessing risk factors has been the relative contribution of genetics and environment to pathogenesis. Comparatively high levels of aluminum and unusually low levels of calcium and magnesium have been found in samples of drinking water and garden soils from Guam and two other high incidence foci of ALS and PDC [23]. The effect of these elemental environments to organism has been investigated by animal experiments, and these experiments demonstrated that the abnormal metabolism of minerals is implicated in the pathogenesis of neurodegenerative disorders [33,34]. It was also revealed in studies on some Guamanian patient [12, 13, 24, 34] that there is prominent accumulation of aluminum and calcium within the nuclear region and surrounding cytoplasm of NFT-bearing hippocampal neurons. The aluminum was reported to yield stable complexes with acids and to produce the accumulation of ammonia that lead to the neuronal death and furthermore it created potential for Fe-induced oxidative stress [35, 36]. But the behavior of transition metal elements in the generation of oxidative stress is poorly understood.

In one study by Shikine et al. (2002), quantitative XRF analysis was performed on substantia nigra (SN) tissues from a patient with Guamanian PDC and a control case. The local distributions and concentrations of intracellular trace metal elements were measured at the single cell level in order to obtain the insight about the neuronal cell death in PDC. The samples from each case were also investigated histologically by staining after XRF analysis.

6.2.2 Sample Preparation

In this study by Shikine et al., the samples were brain tissues obtained by autopsy from a 56-year-old male patient with a Guamanian PDC. Autopsies were performed within 88 hours after death and the whole brain without spinal cord was fixed in 10 % formalin. The samples were prepared as thin sections of 8 μm thickness from tissue blocks dissected from pars compacta of the substantia nigra (SN) and embedded in paraffin. The sections were mounted on a mylar film. The SN tissues of an age-matched male patient who died with non-neurological disorder were treated in the same way and were used as a control. Serial sections adjacent to the section used for X-ray analysis were stained by the hematoxylin-eosin staining in order to verify the distribution and extent of NFT involvement.

6.2.3 Experimental Procedures and Results

The SR-XRF analyses in this study were performed at Photon Factory in beam line 4A. The incident X-ray energy was 14.3 keV and the beam size was approximately $8 \times 5\ \mu m^2$. The detailed set-up of the beam line is described previously. The analyses were carried out in air. The optical microscope observation of the unstained PDC nigral section indicated that the section contained only a few neuromelanin granules, much fewer than the control. The loss of melanized neurons in the SN is characteristic of PDC [37].

The elemental distribution images were obtained in the areas that contained surviving neuromelanin granules. Figures 6.1 and 6.2 show the elemental images of (b) Fe and (c) Zn and the corresponding (a) microscopic photographs (after staining) obtained from the control and PDC cases, respectively. The scale on the right side of the images shows the count of the detector as the fluorescent X-ray intensity. The measurement areas were

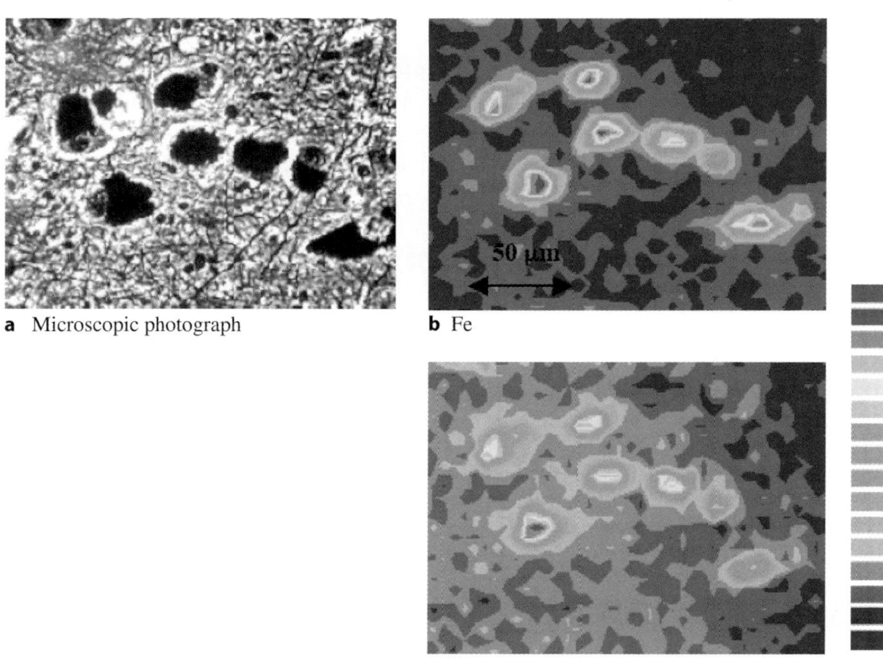

a Microscopic photograph

b Fe

c Zn

Fig. 6.1a–c. The elemental images of (**b**) Fe and (**c**) Zn obtained in the substantia nigra tissue from a control case shown in (**a**) microscopic photograph. Red and blue pixels, respectively, show areas of high and low intensities. It can be seen that the neurons in the control samples contained Fe and Zn

$190 \times 140\,\mu m^2$, $45 \times 70\,\mu m^2$ and $60 \times 60\,\mu m^2$, and the measurement times were 3 s/point, 6 s/point and 7 s/point for Fig. 6.1, 6.2A and 6.2B, respectively. The ranges of intensity were from 0 to 41 for Fe and 0 to 44 for Zn in Fig. 6.1, from 0 to 41 for Fe and 0 to 44 for Zn in Fig. 6.2A and from 0 to 41 for Fe and 0 to 44 for Zn in Fig. 6.2B, respectively.

From the results of the imaging, the measurement points were selected for further point-measurement. The selected points in the control and PDC samples are shown as Fig. 6.3a and 6.3b, which are the optical microscopic photographs after the staining. In the sample from the control case, the points of A1-3 and B1-2 indicate neuromelanin granules and nigral tissues, respectively. XRF spectra were obtained at these points. The typical spectra that were measured in each section are shown in Fig. 6.4a. The measurement time was 200 seconds. Quantitative analysis was then applied to the spectra and the calculated values for concentrations of Fe, Cu, and Zn are shown in Table 6.1a.

The analysis for the sample from the PDC case was performed in the same way as the control case. Measurement points A'1–3, B'1–2 and C'1 in Fig. 6.3b were determined in neuromelanin granules, nigral tissues and glial

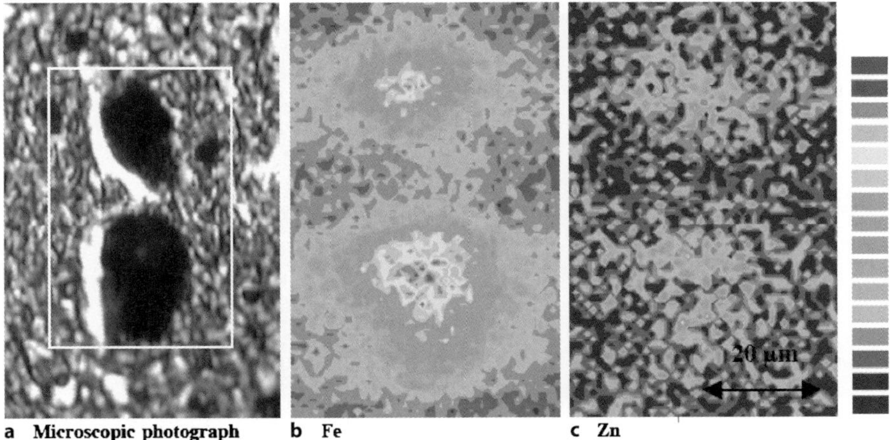

a Microscopic photograph   b Fe   c Zn

Fig. 6.2Aa–c. The elemental images of (**b**) Fe and (**c**) Zn obtained in the substantia nigra tissue from the PDC case shown in (**a**) microscopic photograph. White rectangle indicates the region of imaging. It can be seen that the colony contained Fe while the distribution of Zn was hardly detected due to the low concentration

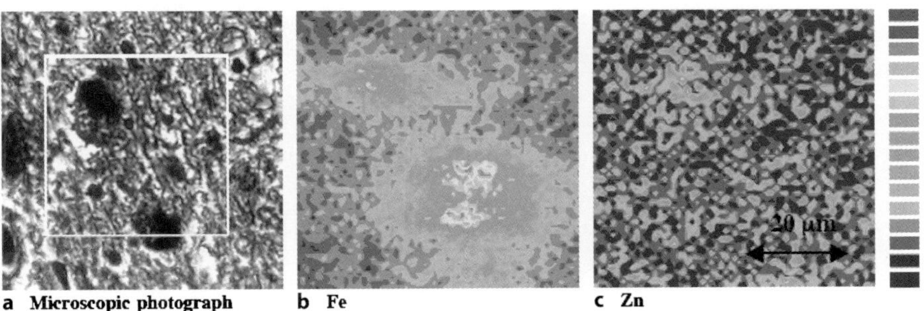

a Microscopic photograph   b Fe   c Zn

Fig. 6.2Ba–c. The elemental images of (**b**) Fe and (**c**) Zn obtained in the substantia nigra tissue from the PDC case shown in (**a**) microscopic photograph. It can be seen that the colony contained Fe while the distribution of Zn was hardly detected due to the low concentration

cells, respectively. The typical XRF spectra obtained in each section from the PDC case are shown in Fig. 6.4b. The concentrations of Fe, Cu, and Zn that were calculated from spectra and the results are also shown in Table 6.1b.

6.2.4 Discussion

Figure 6.1 shows that the neuromelanin granules in the control sample contained more Fe and Zn than the surrounding nigral tissues. The neuromelanin granules in the sample from the PDC case shown in Fig. 6.2A and 6.2B, however, contained Fe, but the Zn contents were not detected because of their low concentrations.

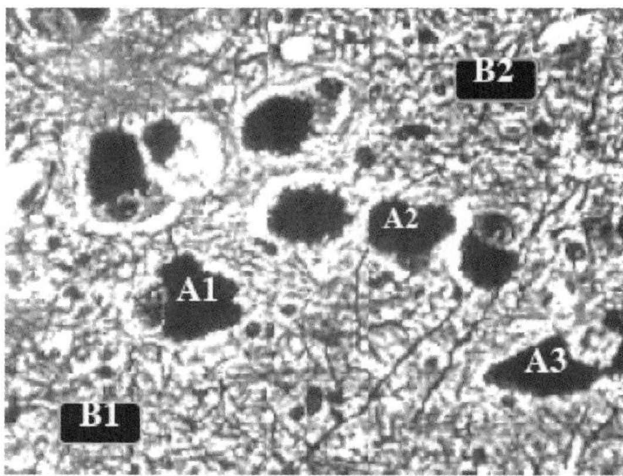

a   **The measurement points in the sample from the control case**

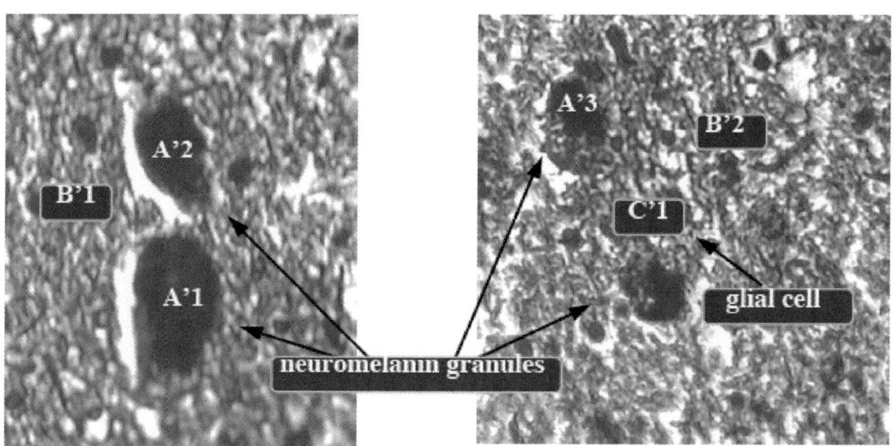

b   **The measurement points in the sample from the PDC case**

Fig. 6.3a,b. The measurement points selected for quantitative XRF analysis are shown with the optical microscopic photographs of (**a**) the control sample and (**b**) the different areas of the sample from the PDC case. The neuromelanin granules, the nigral tissues, and the glial cell are observed at measurement points A (A'), B (B') and C', respectively

This result was also confirmed by XRF spectrum analysis. The spectra obtained in the neuromelanin granules from the control subject (the typical spectra is shown in Fig. 6.4a) showed clear peaks of S, Fe, Ni, Cu and Zn, but in the spectra from the PDC case (the typical one is shown in Fig. 6.4b) Ni and Zn were not detected. On contrary, the peak of As, which was detected in the neuromelanin granules with the PDC case, was not seen in the spectra

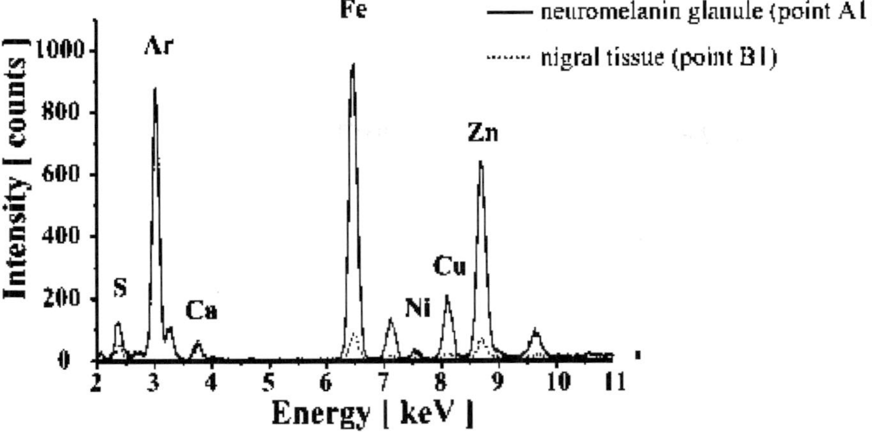

a   Typical XRF spectra obtained in the sample from the control case

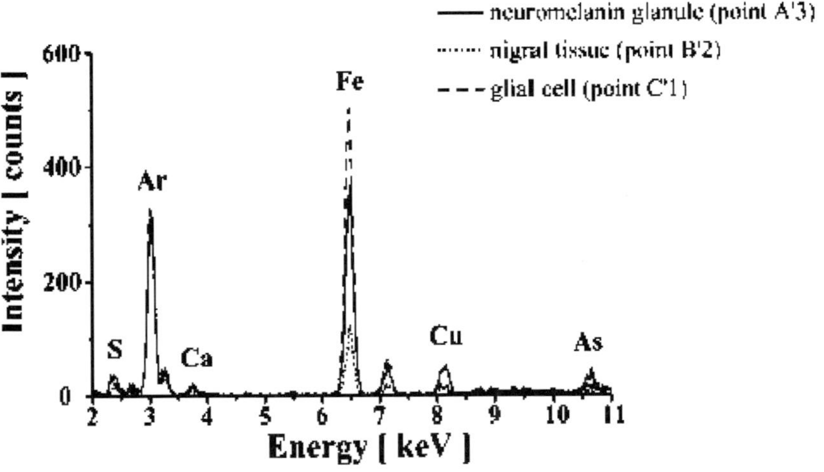

b   Typical XRF spectra obtained in the sample from the PDC case

Fig. 6.4a,b. The typical XRF spectra measured in the substantia nigra (SN) tissues of patients with (**a**) the control and (**b**) the Guamanian PDC cases. Measured points in (**a**) and (**b**) are shown in Fig. 6.3. Points A (A'), B (B') and C' are located in the neuromelanin granules, in the nigral tissues, and in the glial cell, respectively

from the control case. In the spectrum obtained in the glial cell, the high peak of Fe is seen but the peaks of S, Cu, and As are as low as those measured in nigral tissues.

Table 6.1. The quantification results obtained from XRF spectra that were measured of patients with (**a**) the control and (**b**) the Guamanian PDC cases

Measurement points	Concentration [ppm]		
	Fe	Cu	Zn
Neuromelanin granule			
A1	2.2×10^3	2.9×10^2	8.1×10^2
A2	2.0×10^3	1.7×10^2	6.3×10^2
A3	2.4×10^3	1.7×10^2	5.6×10^2
Nigral tissue			
B1	3.1×10^2	9.3×10^1	1.4×10^2
B2	2.0×10^2	3.6×10^1	8.0×10^1

(**a**) The quantification results from the control case

Measurement points	Concentration [ppm]		
	Fe	Cu	Zn
Neuromelanin granule			
A'1	3.1×10^3	3.2×10^2	$< 4.0 \times 10^1$
A'2	2.7×10^3	3.0×10^2	$< 4.0 \times 10^1$
A'3	2.3×10^3	2.1×10^2	$< 4.0 \times 10^1$
Nigral tissue			
B'1	7.3×10^2	6.0×10^1	$< 3.0 \times 10^1$
B'2	7.2×10^2	5.2×10^1	$< 2.0 \times 10^1$
Glial cell			
C'1	3.2×10^3	6.2×10^1	$< 2.0 \times 10^1$

(**b**) The quantification results from the PDC case

Zn is essential for the function of many enzymes and especially important for neurons because it plays a role of preventing apoptosis induced by free radicals as the active center of Zn/Cu superoxide dismutase (Zn/CuSOD). It is possible that the neurons of the patients affected with PDC are subject to the oxidative stress due to the decrease of Zn/CuSOD. Furthermore, it was revealed in the study about ALS that mutated SOD, which may bind the Zn less robustly, undergo reverse catalysis and produce rather than consume superoxide. The function of generating peroxide and triggering tyrosine nitration of mutant SOD was also reported [5]. These results may be the evidence that these phenomena had actually progressed in the substantia nigra of patients with PDC.

It is widely considered that the neuronal cell death in neurodegenerative disorders is triggered by nutritional deficiencies of Ca and Mg leading to secondary hyperparathyroidism that then facilitates the entry of Ca and toxic heavy metals into the brain [38]. Though Ahlskog et al. analyzed the patients' hair, nail and blood, the evidence of accumulation of heavy elements such as Al, As, Cd, Cu, Fe, Pb, Mg, Hg, and Zn was not found. But in this study, the accumulation of As in neuromelanin granules was clearly shown. Arsenic is a strong vascular toxicant and is possibly the direct cause of neuronal cell death [39]. It is, therefore, plausible that the abnormal metabolism was induced and the accumulation of As finally occurred and resulted in neuronal dysfunction.

XRF elemental imaging shown in Fig. 6.2Bb indicates that iron accumulated outside of the neuromelanin granules. Hematoxylin-eosin and Bodian stainings after X-ray analysis revealed that the glial cell was located in this region. The iron had presumably been released from the neuromelanin granule and phagocytosed by the glial cell. McGeer et al. reported that microglia or macrophages had phagocytosed dopaminergic neurons in the SN of postmortem Parkinson's and Alzheimer's disease brains [40, 41]. It was revealed that the similar phagocytosis by glial cells had occurred in the tissues of the PDC afflicted patient.

The results of quantitative XRF analysis shown in Table 6.1 indicated that neuromelanin granules of Parkinsonian substantia nigra (SN) contained higher levels of Fe than those of the control. The concentrations were in the ranges of 2,300–3,100 ppm and 2,000–2,400 ppm, respectively. On the contrary, Zn and Ni in neuromelanin granules of SN tissue from the PDC case were lower than those of the control. Especially Zn was less than 40 ppm in SN tissue from the PDC case while it was 560–810 ppm in the control. The increase of Fe is considered to be closely related to the neurodegeneration and cell death by promoting the generation of free radicals that induce the oxidative stress.

6.2.5 Conclusion

In summary, from the results of XRF spectroscopy Shikine et al. showed that Fe had accumulated in neuromelanin granules, likely due to the progress of PDC. On the contrary, a reduction of Ni and Zn in the neuromelanin granules was found in the PDC case as compared to the control. In neuromelanin granules the concentrations of Fe were in the ranges of 2300–3100 ppm and 2000–2400 ppm in the PDC and the control cases, respectively and for Cu 210–320 ppm and 170–290 ppm, respectively. Zn was hardly detected in the PDC case, but in the control case the concentration was 560–810 ppm. In the PDC case, the glial cell adjacent to neuromelanin granules contained Fe with a high concentration of 3200 ppm. Shikine et al. concluded that the decrease of Zn and accumulation of Fe revealed in the sample from the PDC case may be the direct evidence of the generation of the cytotoxicity such as the oxidative stress and tyrosine nitration. The accumulation of As, which is a strong vascular toxicant, was also confirmed in the study. Phagocytosis of the glial cell that was reported in Parkinson's and Alzheimer's brains was observed.

6.3 Chemical State of Iron in Parkinsonism-Dementia complex (PDC)

6.3.1 Experimental Procedures and Results

In another study, Fujisawa et al. (2002) applied XRF and XANES to the substantia nigra from the same samples, namely post-mortem tissues of a patient with parkinsonism-dementia complex (PDC) and a control subject, and carried out chemical state imaging which differentiates between Fe^{2+} and Fe^{3+} in iron components. This work is discussed in detail in the following. For reference in the XANES analyses, they used FeO (99.9% purity) from High Purity Chemetals Laboratory Ltd. and and Fe_2O_3 powders (98% purity) from Wako Pure Chemical Industries Ltd..

The measurements were performed in vacuum, at beam line 39XU of SPring-8, Japan Synchrotron Radiation Research Institute (JASRI). Monochromatic incident photon beams were restricted by a set of $x - y$ slits and a pinhole to produce about a beam $10\,\mu m$ in diameter.

XRF analyses were performed for elemental mapping. Fluorescent X-rays were collected by a solid state detector (SSD). Fe K-edge XANES analyses were performed in the energy range of 7.100 to 7.160 keV at 0.5eV intervals. The data were measured in fluorescence mode for biological specimens and in transmission mode for the reference samples. Incident and transmitted photon flux was monitored with an air-filled ion chamber. Fe K edge fluorescent X-rays were also collected by a solid state detector (SSD).

In the XANES measurement, the incident energy near the absorption edge is chosen so that selective excitation of specific chemical species will occur. XANES spectra of FeO (Fe^{2+}) and Fe_2O_3 (Fe^{3+}) (reference materials) are shown in Fig. 6.5. At 7.16 keV energy, above the absorption edge, both Fe^{2+} and Fe^{3+} are excited. At 7.12 keV energy, near absorption edge, Fe^{2+} is selectively excited and the excitation of Fe^{3+} is suppressed. Since fluorescent X-rays are emitted accompanied with the excitation of the absorbing elements, XRF imaging that distinguishes between chemical states can be obtained because of the sensitivity of the X-ray absorption coefficient to the chemical state [42, 43].

Fujisawa et al. used the following procedure to obtain chemical state imaging. First, XRF imaging was performed with the incident X-ray energy at 7.160 keV (above the edge) and at 7.120 keV (near the edge). Then the yields of Fe^{2+} and Fe^{3+} were derived using the XRF yields of each point and the absorption coefficients at incident energies of 7.160 keV and 7.120 keV (Table 6.2).

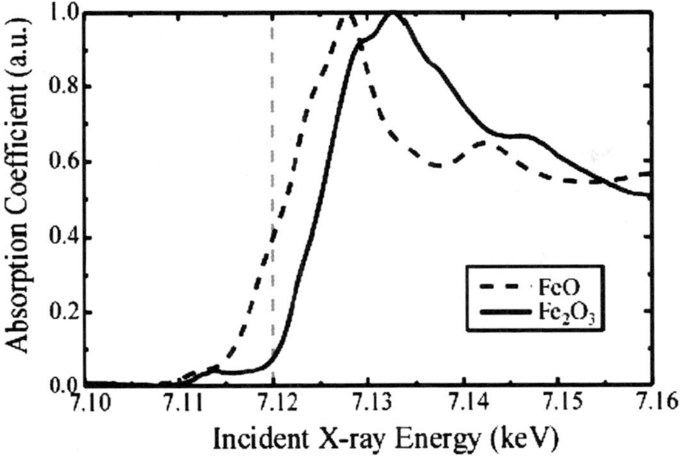

Fig. 6.5. X-ray absorption fine structure (XAFS) spectra of FeO (Fe^{2+}) and Fe_2O_3 (Fe^{3+})

Table 6.2. X-ray absorption coefficient of FeO (Fe^{2+}) and Fe_2O_3 (Fe^{3+}) at the incident energy of 7.160 keV and 7.120 keV

	7.160 keV	7.120 keV
Fe_2O_3 (Fe^{3+})	0.498	0.075
FeO (Fe^{2+})	0.565	0.399

The results of the XRF analyses on substantia nigra tissues of a PDC case and a control case from this work are shown in Fig. 6.6. The optical microscopic photograph of the double-stained tissue (hematoxylin-eosin Bodian to determine cell type and neurofibrillary tangles) and XRF imaging of iron in the PDC tissues are shown in Fig. 6.6a,b, respectively. Two neuromelanin granules released from dead neurons and some glial cells surrounding the neuromelanin are observed. As in Shikine et al. study reported above, iron concentrations are detected in the neuromelanin granules and one of the glial cells. Chemical state imaging which separates Fe^{2+} and Fe^{3+} concentrations was performed in the same area of the PDC tissue (Fig. 6.6c,d). Distributions of Fe^{2+} and Fe^{3+} were well distinguished in the PDC tissue. Fe^{2+} concentration is observed in the neuromelanin, and Fe^{3+} concentration is observed in the glial cell. In the glial cell, Fe^{3+} concentration is at a high density, while the Fe^{2+} is low. This indicates that Fe^{3+} is the predominant valence state of iron contained in the glial cell.

The optical microscopic photograph and XRF imaging of iron in tissue from the control subject are shown in Fig. 6.7a,b, respectively. Several nigral neurons containing neuromelanin can be observed. Iron concentrations were

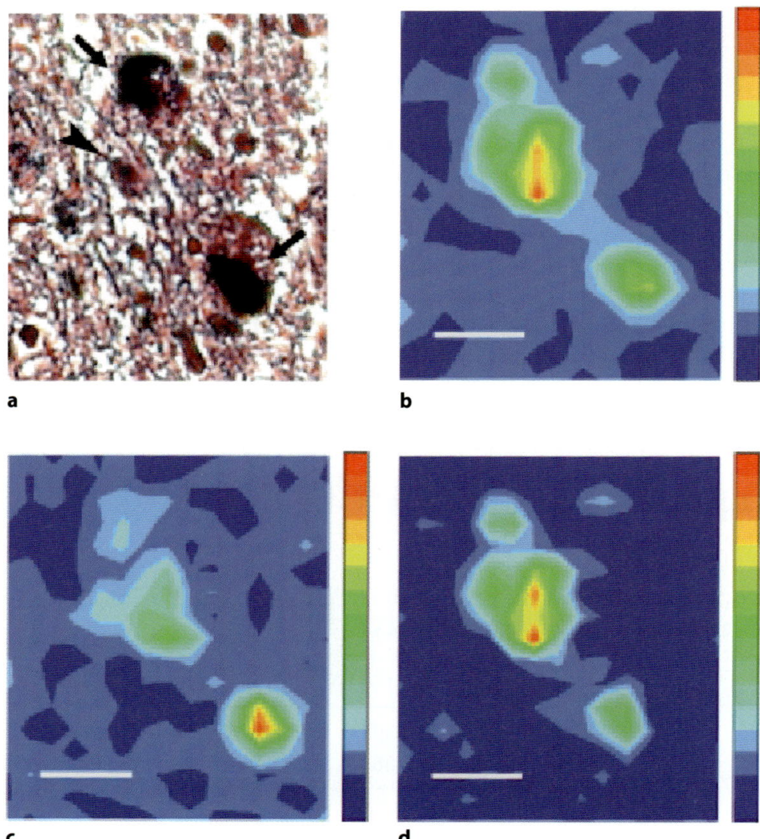

Fig. 6.6a–d. XRF imaging of substantia nigra tissue from a Parkinsonism-Dementia complex case. The measurement area was $70 \times 80\,\mu$m, and was divided into 14×16 pixels. (**a**) Optical microscope photograph of the measurement area taken after X-ray analysis. *Arrows* indicate neuromelanins and an arrowhead indicates a glial cell, where iron accumulation was observed. (**b**) X-ray fluorescence imaging of yield of iron. The range of the fluorescent X-ray intensity is from 0 to 22 counts per second. (**c**) Chemical state imaging of Fe^{2+}. (**d**) Chemical state imaging of Fe^{3+}. The range of the density is from 0 to 40 for Fe^{2+}, from 0 to 60 for Fe^{3+} (arbitrary unit). Scale bars are $20\,\mu$m

readily detected in these neuromelanins. The chemical state imaging in the same area of the control tissue is shown in Fig. 6.7c,d. The distributions of Fe^{2+} and Fe^{3+} in melanized neurons were similar to that of control. Again, in the control tissues, iron components in the melanized neurons were mixed states of Fe^{2+} and Fe^{3+}.

Fe K-edge XANES analyses were applied to selected points where high iron concentrations were detected in the tissues in order to analyze chemical states in detail. Figure 6.8 shows the XANES spectra in the tissues and

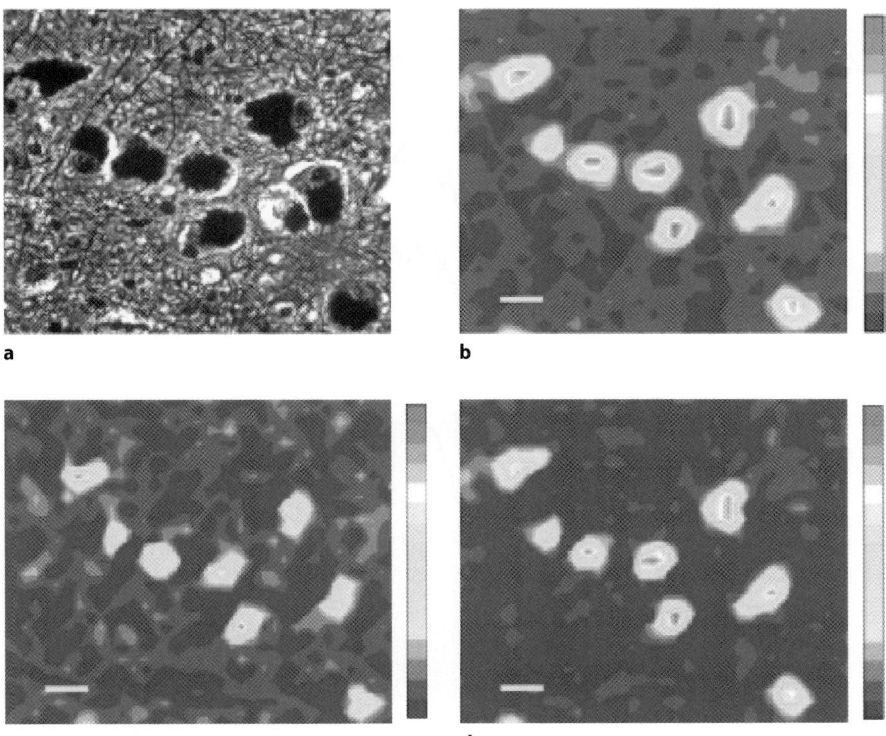

Fig. 6.7a–d. XRF imaging of substantia nigra tissue from a control case. The measurement area was $180 \times 150\,\mu\text{m}$, and was divided into 36×30 pixels. (**a**) Optical microscope photograph of the measurement area taken after X-ray analysis. (**b**) X-ray fluorescence imaging of yield of iron. The range of the fluorescent X-ray intensity is from 0 to 18 counts per second. (**c**) Chemical state imaging of Fe^{2+}. (**d**) Chemical state imaging of Fe^{3+}. The range of the density is from 0 to 25 for Fe^{2+}, from 0 to 70 for Fe^{3+} (arbitrary unit). Scale bars are $20\,\mu\text{m}$

those of reference samples (FeO and Fe_2O_3). Measurement points are in the melanized neuron of control, the neuromelanin granule and the glial cell of the PDC tissue. The ordinate and abscissa represent the absorption coefficient and incident X-ray energy, respectively. The spectra were normalized by the absorption jump, defined as the difference between the highest and the lowest point in each spectrum. The absorption edge is defined at the half-height of the absorption jump.

From spectra of Fig. 6.8, the chemical shifts of iron contained in neuromelanin of the PDC case and the control case are deduced to be 1.1 eV and 1.8 eV, respectively, from that of Fe^{2+} in FeO. For the glial cell of the PDC case the is 3.5 eV. This is close to the chemical shift of Fe^{3+} from Fe_2O_3 of about 3.7 eV. The absorption edge of the iron contained in the neuromelanin within the PDC and the control tissues are mixed states of Fe^{2+} and Fe^{3+}.

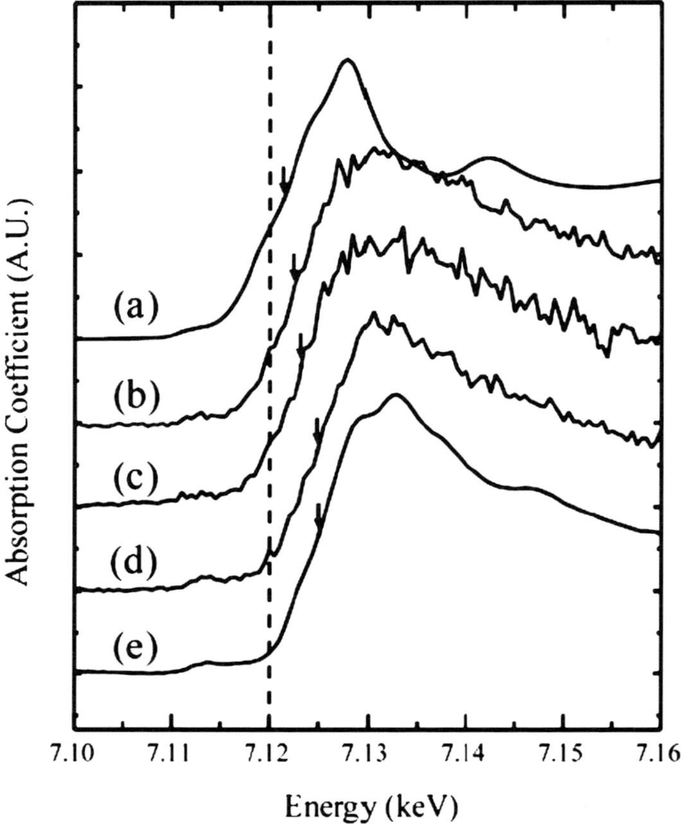

Fig. 6.8. Fe K-edge XANES spectra of (a) FeO powder, (b) the neuromelanin granule from the PDC case, (c) the neurolemanin granule from the control case, (d) one glial cell from the PDC case, and (e) Fe_2O_3 powder. *Arrows* indicate the absorption edge. The incident beam size was about 10 μm in diameter for (b)–(d)

In comparison with the absorption edge of the iron in the neuromelanin of the PDC and the control case, the chemical state of the iron contained in the neuromelanin within the PDC and the control tissue are almost similar. These results are in accordance with the chemical state imaging.

6.3.2 Quantitative Analyses and Fe^{3+}/Fe^{2+} Ratio

Quantitative analysis of the concentrations of iron and other elements contained in the PDC and control tissues was performed. The quantitative analysis was applied to the nigral neurons that have neuromelanin granules, one of glial cells and matrix in the tissue of PDC and control (Table 6.3). The concentrations of iron in the nigral neurons with PDC are $1{,}000 \sim 3{,}700$ ppm, which is higher than those in the control nigral neurons $800 \sim 2{,}000$ ppm.

Table 6.3. Concentrations of the elements in the PDC tissues and the control tissues. The unit is ppm. The density of the tissues is assumed to be $1.0\,\mathrm{g/cm^3}$

		P	S	Cl	Ca	Cr	Mn	Fe
PDC	Neuron	3,168.2	8,806.9	1,027.6	430.9	83.8	142.6	3,754.9
	Neuron	1,917.1	8,476.0	826.4	248.1	52.8	134.5	3,165.6
	Neuron	1,237.0	3,291.5	311.7	175.6	50.8	91.5	1,018.5
	Neuron	1,935.9	6,752.1	661.5	245.7	68.5	112.5	2,621.7
	Glial cell	1,106.9	1,927.7	292.7	167.8	37.5	181.4	3,828.7
Control	Neuron	1,644.7	4,567.2	542.7	405.6	6.7	21.5	962.8
	Neuron	1,865.5	5,577.0	458.9	638.6	20.7	16.5	845.4
	Neuron	2,097.5	5,891.4	640.3	472.3	1.7	12.0	1,487.0
	Neuron	1,940.4	4,258.0	439.3	412.8	14.2	20.8	815.2
	Neuron	1,801.5	4,451.4	630.8	334.3	–	10.3	1,061.6
	Neuron	2,468.3	7,343.7	682.7	505.7	12.3	29.0	1,933.0
	Neuron	1,799.5	5,119.8	469.5	297.3	8.1	12.6	1,166.4
	Neuron	1,247.2	5,531.4	465.8	424.2	7.7	17.1	1,959.2
PDC	Matrix1	809.5	1,835.0	114.6	113.0	–	56.6	365.2
	Matrix2	462.2	1,913.2	139.2	345.5	24.6	47.5	401.0
	Matrix3	458.3	1,294.0	106.5	330.0	20.0	74.0	484.5
Control	Matrix1	1,467.6	3,267.2	264.9	340.4	–	15.5	164.4
	Matrix2	1,815.9	2,660.4	339.3	241.4	8.7	–	305.2

This result indicates that nigral neurons with PDC have iron accumulation compared to those with control. The increase of Cr and Mn also can be seen in PDC nigral cells.

In order to obtain the Fe^{3+}/Fe^{2+} ratio of iron contained in the tissues, XRF analysis was performed at several points in PDC nigral tissues of a patient at two incident X-ray energies of $7.160\,\mathrm{keV}$ and $7.120\,\mathrm{keV}$. An optical microscopic photograph of the tissue is shown in Fig. 6.6, where some neuromelanin granules and glial cells can be seen. Typical XRF spectra in the tissue are shown in Fig. 6.9. The ordinate and abscissa represent XRF intensity and fluorescent energy, respectively. XRF intensity is normalized using fluorescent intensity divided by I_0. I_0 is the incident X-ray intensity, propotional to photons per second, measured by ionized chamber. Table 6.4 shows the XRF yield of iron at several points in the tissue where iron was detected with high density, at neuromelanin granules and one of the glial cells. The Fe^{3+}/Fe^{2+} ratio is obtained using XFR yield and absorption coefficient (Table 6.2). We can see that Fe^{3+}/Fe^{2+} ratio varies at the points in the tissues. For example, Fe^{3+}/Fe^{2+} ratio of iron contained in the glial cell is about tenfold higher than that in the neuromelanin granule (A). Chemical state of iron can be well distinguished at the different points in the tissues.

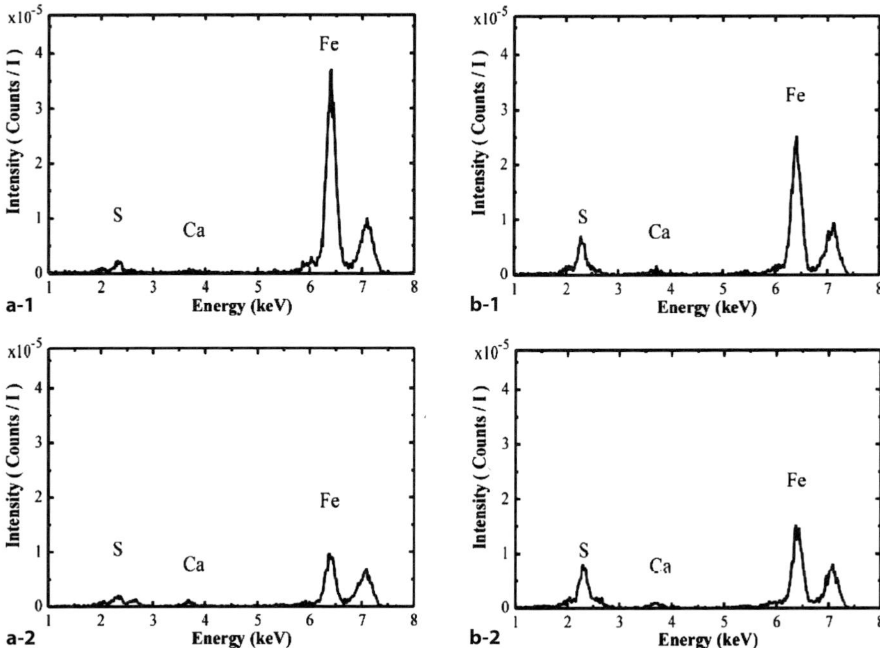

Fig. 6.9a-1–b-2. Typical XRF spectra in the tissue with PDC. Measurement points were inside of glial cell and neuromelanin granules. (**a-1**) Glial cell, incident energy of 7.160 keV. (**a-2**) Glial cell, 7.120 keV. (**b-1**) Neuromelanin granules, 7.160 keV. (**b-2**) Neuromelanin granules, 7.120 keV. Measurement time was 200 s for each point

Table 6.4. Peak area of X-ray fluorescence spectra in the tissues of PDC with the incident energy of 7.160 keV and 7.120 keV. Measurement time was 200 seconds for each point. XRF yields was normalized by incident X-ray intensity

Target	XRF yield (7.160 keV)	XRF yield (7.120 keV)	Fe^{3+}/Fe^{2+}
Glial cell	$3.98E-05$	$9.81E-06$	5.46
Neuromelanin(A)	$2.60E-05$	$1.42E-05$	0.47
Neuromelanin(B)	$1.03E-05$	$2.94E-06$	3.55
Extra cellar tissue	$5.17E-06$	$1.90E-06$	1.77

6.3.3 Discussions and Summary

In this work chemical state analyses were applied to pathological tissues. Chemical state imaging which separates Fe^{2+} and Fe^{3+} was obtained with the following assumptions: 1) the absorption coefficient curve of different valence states of iron (Fe^{2+} and Fe^{3+}) was represented by only iron oxide

(FeO and Fe_2O_3), 2) iron contained in the tissues is a superposition of Fe^{2+} and Fe^{3+}. It should be noted that these assumptions are too simplified to represent the complex system, which contains chlorine and other compounds of Fe. These assumptions may be justified considering the important role of oxides of iron in oxidation and reduction activities of brain. Under these assumptions, the chemical states of iron contained in the tissues were well distinguished.

XRF analyses revealed an excessive accumulation of iron in the neuromelanin granules of the PDC tissues. This result is in good agreement with previous studies which show iron concentration in neuromelanin in tissues from PD cases [28, 44]. In this study, iron contained in the neuromelanin of the PDC and the control cases are mixed states of Fe^{2+} and Fe^{3+}. In a patient with PD, selective vulnerability can be seen in melanized neurons [45, 46], so there are several hypotheses that suggest that neuromelanin may be part of what is responsible for neuronal degeneration [47]. The chemical state of the iron in dopaminergic neurons is important in understanding the role of iron in the pathogenesis.

In this study, high accumulation of iron can be seen in one of glial cells near the neuromelanin granules in the PDC tissue. In this area, the chemical state of iron has shifted toward Fe^{3+}. In the previous study [26], high accumulation and chemical shifts of iron were found in some of the glial cells of a PD case. McGeer reported that numerous microglia or macrophages phagocytosed dopaminergic neurons in SN pars compacta of post-mortem parkinsonian brains [40, 41]. This can be related to excessive accumulation of iron, specially, iron in the higher charge state (Fe^{3+}). These results (Fujisawa et al. 2002) also show excessive accumulation and significant chemical shifts of iron in one of the glial cells. This can be related to an important aspect of the role of iron in glial cells of PDC and PD cases.

The XANES spectra of iron contained in the tissues were not smooth. In the beam and tissue interaction region, an estimate of the physical parameters showed that the volume the tissues was about $600\,\mu m^3$ and weight was $600\,pg$ ($pg = 10^{-12}\,g$), and the iron content in the tissue was about $2.5\,pg$ [48]. A typical value for the detection limit [22] obtained by using the signal/noise ratio in a typical spectrum was $0.1\,pg$. The XANES spectrometry using a microbeam was performed for the iron contents of a few pg or less. The practical measurement time at each point was about 50 to 60 seconds, and the total measurement time for each spectrum was about 2 hours. The fluctuation in the XANES spectra of trace elements in the tissues is inevitable when using SR microbeams for practical applications.

In this study chemical state imaging of iron could clearly discriminate between tissues containing mainly Fe^{2+} and tissues containing Fe^{3+} from the PDC case and the control. This method can be used to help us understand the behavior of transition metals inside and outside cells, and can be used widely for investigations in neurology, basic cell biology and pathological fields.

6.3.4 Conclusion

In this chapter, application of SR methods demonstrated a novel approach to the study of biological function of cells through the investigation of the intracellular elemental conditions with high spatial resolution and sensitivity. Micro-XRF and XANES analyses are ideal techniques for providing the distributions, concentrations and chemical states of trace elements. Furthermore, their non-destructive nature allows histological analysis of the samples by staining after the elemental analysis. These features cannot be obtained by the conventional techniques and were exploited in the present study. The analyses enable the absolute concentrations of a wide range of the trace elements to be determined. This quantitative information is of significant importance for elucidating the role of intracellular trace elements in the investigation of biomedical samples.

The case study of parkinsonism-dementia complex (PDC) revealed that the substantia nigral tissue obtained from the PDC case showed the decrease of Zn and accumulation of Fe. These may be the direct evidence of cytotoxicity from oxidative stress and tyrosine nitration. The accumulation of As contents that is strongly toxic for tissues was also found in the PDC case. It was clarified that the phagocytosis by the glial cell had occurred.

There is no doubt that micro-XRF and XANES analyses will provide unprecedented opportunities to analyze biological samples in future studies as well as the present work. The quantitative information at this level is not accessible with more traditional imaging and spectroscopic methods and will contribute to the development of therapy of neurodegenerative disorders.

References

1. B. Halliwell, *J. Neurochem.*, **1992**, 59, 1609.
2. S. Fahn, G. Cohen, *Ann. Neurol.*, **1992**, 32, 804.
3. G. Multhaup, A. Schlicksupp, L. Hesse, D. Beher, T. Ruppert, C.L. Masters, K. Beyreuther, *Science*, **1996**, 271, 1406.
4. C.W. Olanow, G.W. Arendash, *Curr. Opin. Neurol.*, **1994**, 7, 548.
5. D.W. Cleveland, J. Liu, *Nature Med.*, **2000**, 6, 1320.
6. E.C. Hirsh, J.P. Brandet, P. Galle, Y. Agid, *J. Neurochem.*, **1991**, 56, 446.
7. E. Sofic, W. Paulus, K. Jellinger, P. Riederer P, M.B. Youdim, *J. Neurochem.*, **1991**, 56, 978.
8. K. Jellinger, E. Kienzl, G. Rumpelmair, P. Riederer, H. Stachelberger, D. Benshachar, M.B. Youdim, *J. Neurochem.*, **1992**, 59, 1168.
9. P.F. Good, C.W. Olanow, D.P. Perl, *Brain Res.*, **1992**, 593, 343.
10. J.R. Connor, B.S. Snyder, J.L. Beard, R.E. Fine, E.J. Mufson, *J. Neurosci. Res.*, **1992**, 31, 327.
11. P.F. Good, D.P. Perl, L.M. Bierer, J. Schmeidler, *Ann. Neurol.*, **1992**, 31, 286.
12. D.P. Perl, D.C. Gajdusek, M.G. Garruto, R.T. Yanagihara, C.J. Gibbs Jr., *Science*, **1982**, 217, 1053.

13. R.M. Garruto, R. Fukatsu, R. Yanagihara, D.C. Gajdusek, G. Hook, C.E. Fiori, *Proc. Natl. Acad. Sci. USA*, **1984**, 81, 1875.

14. M. Gerlach, D. Ben-Shachar, P. Riederer, M.B. Youdim, *J. Neurochem.*, **1994**, 63, 793.

15. J.R. Connor, *"Evidence for iron mismanagement in the brain in neurological disorders, in Metals And Oxidative Damage In Neurological Disorders"*, **1997**, Plenum Press, 23.

16. S.J. Stohs, D. Bagchi, *Free Radic. Biol. Med.*, **1995**, 18, 321.

17. A.I. Bush, *Curr. Opin. Chem. Biol.*, **2000**, 4, 184.

18. P.S.P. Thong, F. Watt, D. Ponraj, S.K. Leong, Y. He, T.K.Y. Lee, *Nucl. Instr. Meth. Phys. Res. B*, **1999**, 158, 349.

19. J.T. Coyle, P. Puttfarcken, *Science*, **1993**, 262, 689.

20. E. Kienzl, L. Puchinger, K. Jellinger, W. Linert, H. Stachelberger, R.F. Jameson, *J. Neurol. Sci.*, **1995**, 134, S69.

21. F. Watt, *Cell. Mol. Biol.*, **1996**, 42, 17.

22. C.J. Sparks Jr., *"Synchrotron Radiation Research"*, **1980**, ed. H. Winick and S. Doniach, Plenum Press.

23. R.M. Garruto, Y. Yase, *Trends Neurosci.*, **1986**, 9, 368.

24. P. Piccardo, R. Yanagihara, R.M. Garruto, C.J. Gibbs, D.C. Gajdusek, *Acta Neuropathol.*, **1988**, 77, 1.

25. G.R. Lachance, F. Claisse, *"Quantitative X-ray fluorescence analysis"*, **1995**, ed. J. Wiley & Sons.

26. S. Yoshida, A.M. Ektessabi, S. Fujisawa, *J. Synchr. Rad.*, **2001**, 8, 998.

27. A.M. Ektessabi, S. Fujisawa, K. Takada, K. Yoshida, H. Maruyama, R.W. Shin, *Int. J. PIXE*, **1999**, 9, 297.

28. A.M. Ektessabi, S. Yoshida, K. Takada, *X-ray Spectrom.*, **1999**, 28, 456.

29. A. Bianconi, A. Congiu-Castellano, P.J. Durham, S.S. Hasnain, S. Phillips, *Nature*, **1985**, 318, 685.

30. S. Della Longa, S. Pin, R. Cortès, A.V. Soldatov, B. Alpert, *Biophys. J.*, **1998**, 75, 3154.

31. A.J. Kropf, B.A. Bunker, M. Eisner, S.C. Moss, L. Zecca, A. Stroppolo, P.R. Crippa, *Biophys. J.*, **1998**, 75, 3135.

32. P.D. Griffiths, B.R. Dobson, G.R. Jones, D.T. Clarke, *Brain*, **1999**, 122, 667.

33. T. Kihira, S. Yoshida, J. Komoto, I. Wakayama, Y. Yase, *Neurotoxicol.*, **1995**, 16, 413.

34. M. Yasui, Y. Yase, K. Ota, *J. Neurol. Sci.*, **1991**, 105, 206.

35. C.X. Xie, M.P. Mattson, M.A. Lovell, R.A. Yokel, *Brain Res.*, 1996, 743, 271.

36. R. Deloncle, O. Guillard, *Neurochem. Res.*, **1990**, 15, 1239.

37. S. Goto, A. Hirano, S. Matsumoto, *Ann. Neurol.*, **1990**, 27, 520.

38. J.E. Ahlskog, S.C. Waring, L.T. Kurland, R.C. Petersen, T.P. Moyer, W.S. Harmsen, D.M. Maraganore, P.C. Obrien, C. Estebansantillan, V. Bush, *Neurology*, **1995**, 45, 1340.

39. W. Zheng, *Microsc. Res. Tech.*, **2001**, 52, 89.

40. P.L. McGeer, S. Itagaki, B.E. Boyes, E.G. McGeer, *Neurology*, **1988**, 38, 1285.

41. P.L. McGeer, S. Itagaki, H. Akiyama, E.G. McGeer, *Ann. Neurol.*, **1988**, 24, 574.

42. K. Sakurai, A. Iida, Y. Gohshi, *Adv. in X-ray Anal.*, **1989**, 32, 167.

43. H. Ade, X. Zhang, S. Cameron, C. Costello, J. Kirz, S. Williams, *Science*, **1992**, 258, 972.

44. L. Zecca, T. Shima, A. Stroppolo, C. Goj, G.A. Battiston, R. Gerbasi, T. Sarna, H.M. Swartz, *Neuroscience*, **1996**, 73, 407.
45. E.C. Hirsh, A. Graybiel, Y. Agid, *Nature*, **1988**, 334, 345.
46. W.R. Gibb, *Brain Res.*, **1992**, 581, 283.
47. A. Kastner, E.C. Hirsh, Q. Jejeune, F. Javoy-Agid, O. Rascol, Y. Agid, *J. Neurochem.*, **1992**, 59, 1080.
48. A.M. Ektessabi, S. Shikine, S. Yoshida, *"Application of Accelerators in Research and Industry" – Sixteenth Int'l Conf.*, ed. J.L. Duggan and I.L. Morgan, **2001**, 720.

7

Investigation
of Neurodegenerative Disorders (II)

7.1 Introduction

In the previous chapter, we introduced the important role played by transition metal elements in many neurodegenerative disorders and concentrated on the case of Guamanian parkinsonism-dementia complex (PDC). Synchrotron radiation-based micro-XRF and XANES analyses are ideal techniques for providing the distributions, concentrations and chemical states of trace elements. Furthermore, their non-destructiveness allows histological analysis of the samples after the elemental analysis. These features cannot be obtained by conventional techniques and represent a new approach to biomedical studies by providing quantitative information which may be of great significance in elucidating the role of intracellular trace elements. In this chapter we extend the application of SR to cases of amyotrophic lateral sclerosis (ALS), Alzheimer's disease and Friedreich Ataxia.

7.2 Amyotrophic Lateral Sclerosis

7.2.1 Introduction

Amyotrophic lateral sclerosis (ALS) is an adult onset disorder characterized by progressive weakness and spasticity, typically leading to death within 3 to 5 years. The pathological features of this disease are the degeneration of large motor neurons in the cerebral cortex, brainstem, and spinal cord and the abnormal accumulation of neurofibrillary tangles (NFT) in the cell body and proximal axon of motor neurons [1, 2]. Between 5 and 10% of cases of ALS are familial ALS (FALS), while others are considered to be sporadic ALS (SALS) with no identifiable genetic or environmental risk factors.

There is consensus that the oxidative stress promoted by reactive oxygen species is deeply related to the neural cell death in the neurodegenerative disorders such as Parkinson's disease, Alzheimer's disease and ALS. Transition metals in the organism like Fe are thought to promote the generation of free radicals such as O_2^-, and to cause oxidative stress. On the other hand, there are enzymes called super-oxide dismutase (SOD) that dissociates free radicals. Cu/Zn SOD, which is the major SOD in human bodies, utilize coordinated Cu and Zn to dissociate free radicals. This enzyme is encoded by the SOD1 gene and converts O_2^- into H_2O_2, which is then metabolized by gluathion peroxidase [3]. The generally accepted mechanism of dismutation involves cyclic reduction and reoxidation of Cu(II) and Cu(I), respectively by single molecules of superoxide [4].

$$O_2^- + \text{Enz-Cu}^{2+} \rightarrow O_2 + \text{Enz-Cu}^+$$
$$\text{Enz-Cu}^+ + O_2^- + 2H^+ \rightarrow \text{Enz-Cu}^{2+} + H_2O_2$$

Recently a series of mutations in a Cu/Zn SOD located on chromosome 21 have been identified in 15 to 20% of FALS cases and 5% of SALS cases [1,5]. It is generally accepted that the mutant Cu/ZnSOD gains novel, cytotoxic activity rather than the loss of dismutase function leads to the generation of oxidative stress [6]. This consideration is based on several experimental results. The typical evidences are that neither elimination nor elevation of wild-type SOD1 was found to affect disease induced by mutant SOD in mice [7] and transgenic mice over-expressing mutant SOD1 protein develop and ALS phenotype [8]. The possible explanation is that the copper in the active center of the mutant molecule becomes accessible to substrates other than O_2^-, resulting in the generation of free radicals and thus in oxidative stress to the cell. The other hypotheses is that mutant proteins misfold and form intracellular aggregates, which may sequester other protein components, reduce the availability of protein-folding chaperones to catalyze folding of other proteins, or reduce proteosome activity needed for normal protein turnover [4].

ALS-SOD can scavenge superoxide as efficiently as wild-type SOD when they contain their full complement of Cu and Zn. The mutations, however, are reported to decrease Zn affinity of wild-type SOD [9]. A4V (substitution of alanine at position 4 by valine), which is the most common SOD mutation, yield SODs with the weakest affinity for Zn and causes rapid disease progression [10]. And it was revealed that the loss of Zn from SODs was sufficient to induce apoptosis in cultured motor neurons. Toxicity required that Cu be bound to SOD and depended on endogenous production of nitric oxide.

The mechanism of ALS is extremely complicated as seen above, so the pathogenesis remains poorly understood, and it has yet to be elucidated how SOD1 mutants cause ALS [4]. Furthermore evidence for oxidative stress is not only found in mutant SOD-related FALS, but also in SALS; it is tempting to speculate that a similar mechanism is at work in both forms of the disease. But it is obvious that transition metals are related to both the generation and dissociation of the cytotoxicity and play an important role in this disease.

Shikine et al. (2002) conducted a study with the aim of investigating how the concentration and the distribution of the metal elements had changed due to the progress of FALS and SALS. Most of past knowledge of this topic described above was obtained from cultured cell or model animals, and the phenomena that actually occurred in the tissues of affected patients were not investigated enough. The quantitative information of trace metal elements in the afflicted motor neurons is expected to bring new insight into the mechanism of cell death in ALS. The study also includes samples from an experiment where cultured mouse cells were injected with ALS DNA in order to confirm the results and conclusion obtained from the investigation of the human tissues.

7.2.2 Sample Preparation

7.2.2.1 Anterior Horn Tissues from FALS and SALS Cases

Shikine et al. used samples of the anterior horn tissues of the spinal cord obtained by autopsy from a control, familial and sporadic ALS cases. After being fixed in 10% formalin, they were embedded in paraffin and were cut into 8 µm thin sections. The sample for XRF analysis was made by putting the section on a mylar film. Serial sections adjacent to the section used for X-ray analysis were stained by the hematoxylin-eosin staining in order to verify the distribution and extent of NFT involvement. Figure 7.1a shows the motor neurons in a stained control sample. In the samples from ALS, the motor neurons had decreased and NFTs had accumulated in several neurons as seen in Fig. 7.1b. This neurofilament disruption and SOD1-induced toxicity are considered to act together [11].

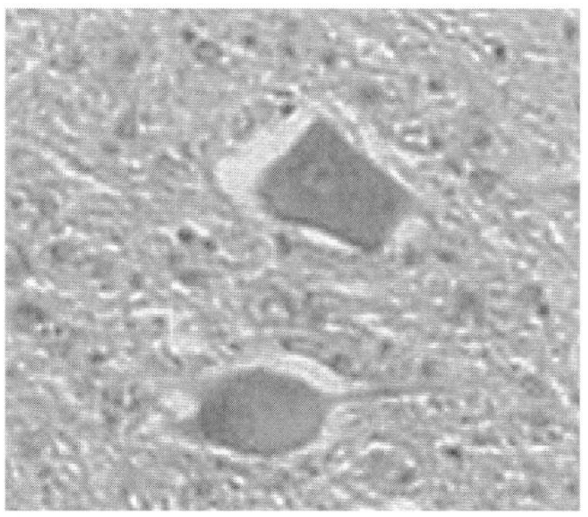

a   Control case

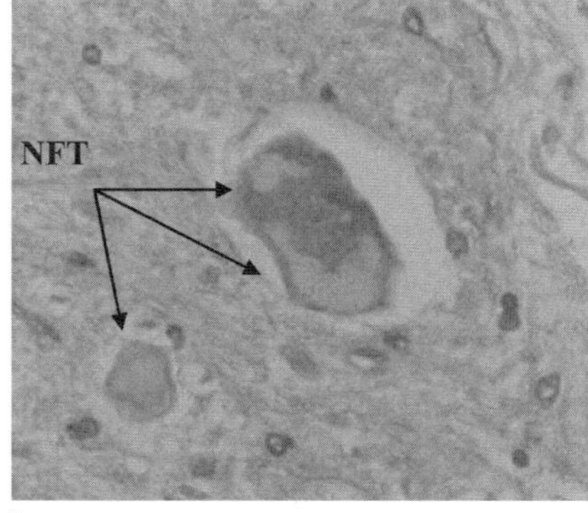

b   FALS case

Fig. 7.1a,b. The optical microscopic photographs of motor neurons in anterior horn tissues from (**a**) a control case and (**b**) the familial amyotrophic lateral sclerosis (FALS) case. The accumulation of neurofibrillary tangles (NFTs) can be seen in (**b**). Picture (**b**) was quoted from [67]

7.2.2.2 Cultured Mouse Cells Injected with ALS DNA

To confirm the experimental result obtained from the investigation of the human tissues, Shikine et al. conducted an experiment where a cultured mouse was injected with wild-type and mutated human SOD1 gene. The main purpose was to analyze the change in the concentrations and chemical states between the cells expressing wild-type SOD and mutant SOD. If the Zn affinity is deteriorated in mutant SOD, the differences are observed in the chemical structures of the Cu and Zn binding site. Though the chemical structures of metal binding sites in human and bovine SOD were analyzed by XANES analysis and X-ray crystallography [12,13], the alteration due to the mutation have not been investigated yet.

The DNA segment coding for a full-length SOD1 with/without mutations and a green fluorescent protein (GFP) was introduced into an expression vector. The resultant plasmid carried wild-type, Ala4Thr (substitution of alanine at position 4 by threonine: A4T) and 2 base-pair deletions in the 126th codon (d126p) mutations. Mutant A4T and d126p, as well as wild-type, SOD1 were expressed in mouse neuroblastome cell, Neuro2a. A4T and d126p mutations were identified in FALS cases [14,15]. As a control sample, the Neuro2a cells injected with only GFP were investigated as well as those injected with SOD1. Figure 7.2 shows the fluorescent microscopic photographs of the cells in which the SOD1 gene were expressed.

These cells were provided by Dr. Ryoichi Nakano, Department of Neurology, Brain Research Institute, Niigata University. His group revealed that the granular cytoplasmic aggregates were formed accompanied by collapse of the cytoplasm in cells expressing mutant SODs, but not in cells expressing wild-type SODs [16,17].

7.2.3 Experimental Procedures and Results

7.2.3.1 Anterior Horn Tissues from FALS and SALS Cases

The SR-XRF analyses in this investigation were performed at Photon Factory in beam line 4A. The incident X-ray energy was 14.3 keV and the beam size was approximately $7 \times 5\,\mu m^2$. The detailed set-up of the beam line is described in Chap. 2. The analyses were carried out in air.

The distributions of trace metal elements at the single cell level were obtained in the control, FALS and SALS samples, respectively. The elemental images of (a) Fe, (b) Cu and (c) Zn are shown in Figs. 7.3, 7.4 and 7.5, respectively. The scale on the right side of the images shows the count of the X-ray intensity. The measurement areas were $120 \times 120\,\mu m^2$, $99 \times 120\,\mu m^2$ and $99 \times 99\,\mu m^2$ and the measurement times were 6 s/point, 6 s/point and 5 s/point respectively. The range of intensity was from 4 to 70 for Fe, 9 to 117 for Cu and 2 to 31 for Zn in Fig. 7.3, from 2 to 48 for Fe, 1 to 24 for Cu and 4 to 551 for Zn in Fig. 7.4 and from 12 to 176

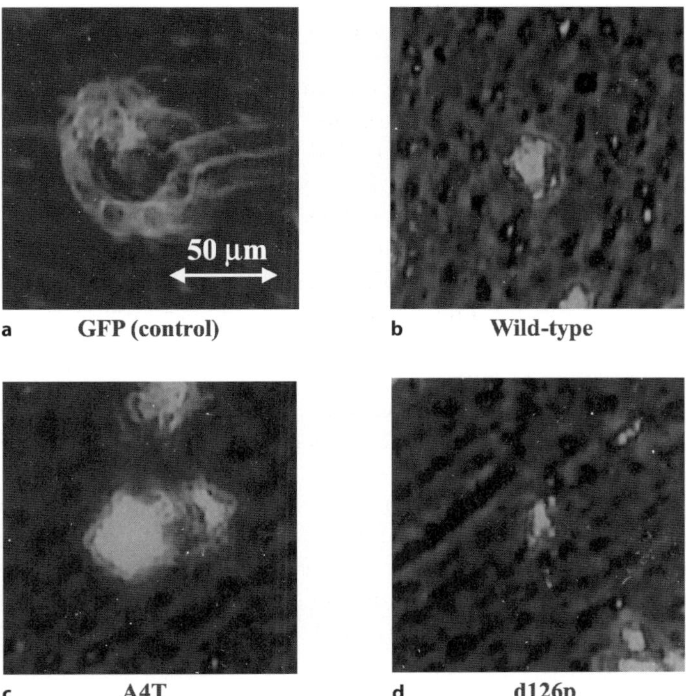

Fig. 7.2a–d. The fluorescent microscopic photographs of the cells in which the SOD1 gene were expressed: (**a**) green fluorescent proteins (GFP), (**b**) wild-type human SOD1, (**c**) mutant A4T and (**d**) d126p SODs were expressed in mouse neuroblastome cell, Neuro2a

for Fe, 0 to 111 for Cu and 0 to 39 for Zn in Fig. 7.5, respectively. From the results of the imaging, the measurement points were selected for further quantitative point-measurement. XRF spectra were obtained at these points to reveal the concentrations of intracellular trace elements. The measurement time was 200 seconds. The typical spectra that were obtained in the motor neurons in the sample from the control, FALS and SALS cases are shown as the solid, broken and dotted lines in Fig. 7.6, respectively. Each spectrum is normalized by the incident X-ray intensity. Quantitative analysis was then applied to all measured spectra and the calculated values for concentrations of transition metal elements such as Fe, Cu and Zn are shown in (a) the table and (b) the bar graph in Fig. 7.7.

7.2.3.2 Cultured Mouse Cells Injected with ALS DNA

The SR-XRF elemental imaging and quantitative analysis in this investigation was performed at Photon Factory in beam line 4A. The incident X-ray

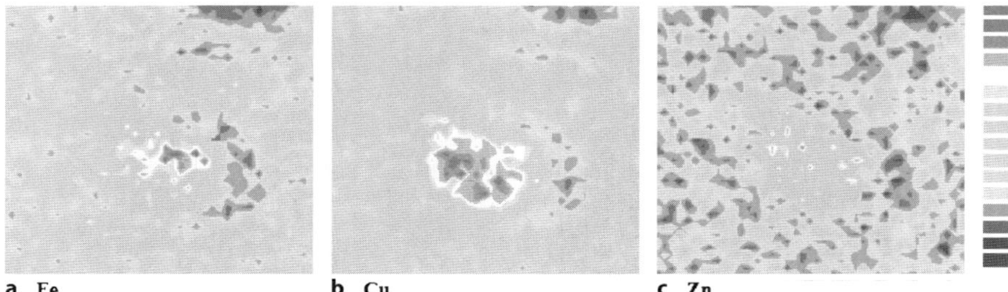

Fig. 7.3a–c. The elemental images of (**a**) Fe, (**b**) Cu and (**c**) Zn obtained in the motor neuron from the control case. Measurement area was $120 \times 120\,\mu m^2$ and the measurement time was 6 s/point. The scale on the right side of the images shows the count of the detector as the fluorescent X-ray intensity. Red and blue pixels, respectively, show areas of high and low intensities. The motor neuron in the control samples contained Fe, Cu and Zn

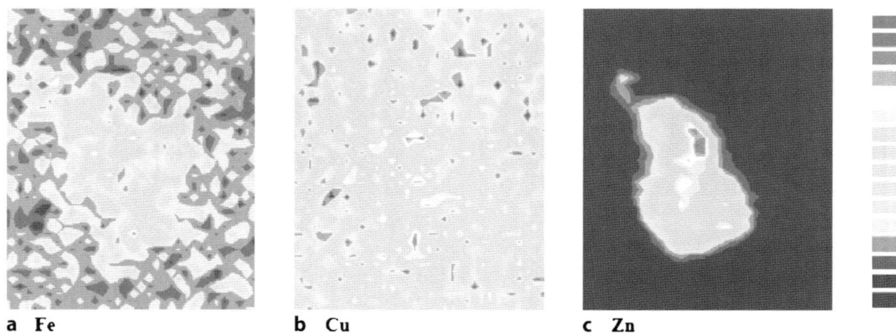

Fig. 7.4a–c. The elemental images of (**a**) Fe, (**b**) Cu and (**c**) Zn obtained in the motor neuron from the familial amyotrophic lateral sclerosis (FALS) case. Measurement area was $99 \times 120\,\mu m^2$ and the measurement time was 6 s/point. The motor neuron in the sample from the FALS case contained Fe and Zn but the concentration of Cu was low

energy was 14.3 keV and the beam size was approximately $7 \times 5\,\mu m^2$. The analysis was carried out in air.

The elemental distributions of P, S, Fe, Cu and Zn were obtained in the cells in which GFP, wild-type, mutant A4T and d126p SODs were expressed. The elemental distributions of (b) P, (c) S, (d) Fe, (e) Cu and (f) Zn in the cells that expressed wild-type SOD1 are shown with (a) the fluorescent microscopic photograph in Fig. 7.8. The measurement area was $81 \times 72\,\mu m^2$ and the measurement times were 8 s/point. The range of intensity was from 0 to 57 for P, 1 to 64 for S, 12 to 50 for Fe, 12 to 47 for Cu and 6 to 77 for Zn. (The results of elemental imaging about the cells that expressed GFP, mutant A4T and d126p are not shown.) From the results of the imaging, the measurement points were selected for further quantitative point-measurement. XRF spectra were obtained at 10 points in each sample to reveal the con-

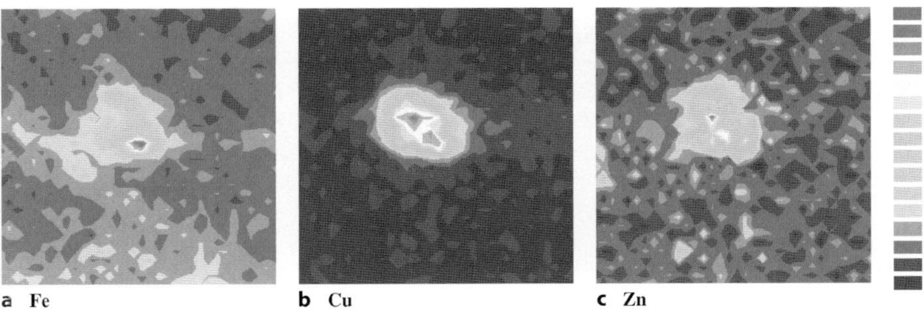

Fig. 7.5a–c. The elemental images of (a) Fe, (b) Cu and (c) Zn obtained in the motor neuron from the sporadic form of amyotrophic lateral sclerosis, (SALS) case. Measurement area was $99 \times 99\,\mu m^2$ and the measurement time was $5\,s$/point. Intracellular Fe, Cu and Zn of the motor neuron from the FALS case were detected at the relatively high intensity

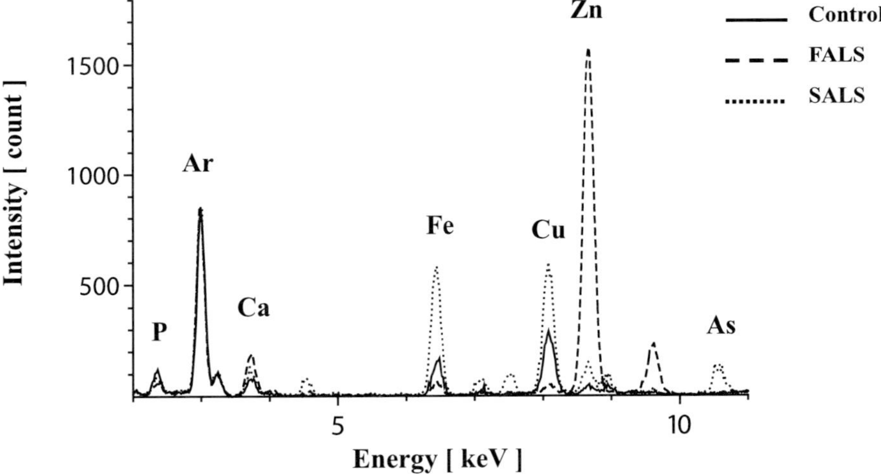

Fig. 7.6. The typical XRF spectra measured in the motor neurons from (a) the control, (b) the FALS and (c) SALS cases. The measurement time was $200\,s$ at each point. The incident beam energy was $14.3\,keV$. Each spectrum is normalized by the incident X-ray intensity

centrations of trace elements. The measurement time was 200 seconds. The typical spectra that were obtained in the cells that expressed (a) GFP, (b) wild-type, (c) mutant A4T and (d) d126p are shown in Fig. 7.9. Each spectrum is standardized with the incident X-ray intensity. Quantitative analysis was then applied to all measured spectra and the calculated values for concentrations of P, S, Cl, Fe and Zn are shown in Table 7.1. The thickness and the density of the samples are assumed as $10\,\mu m$ and $1.0\,g/cm^3$, respectively.

	Fe	Cu	Zn
Control	283.8 ± 89.1	292.9 ± 22.8	40.9 ± 7.4
FALS	109.7 ± 24.5	39.1 ± 8.1	$1,250.3 \pm 407.7$
SALS	845.4 ± 182.3	588.1 ± 106.9	123.8 ± 23.3

a   [ppm]

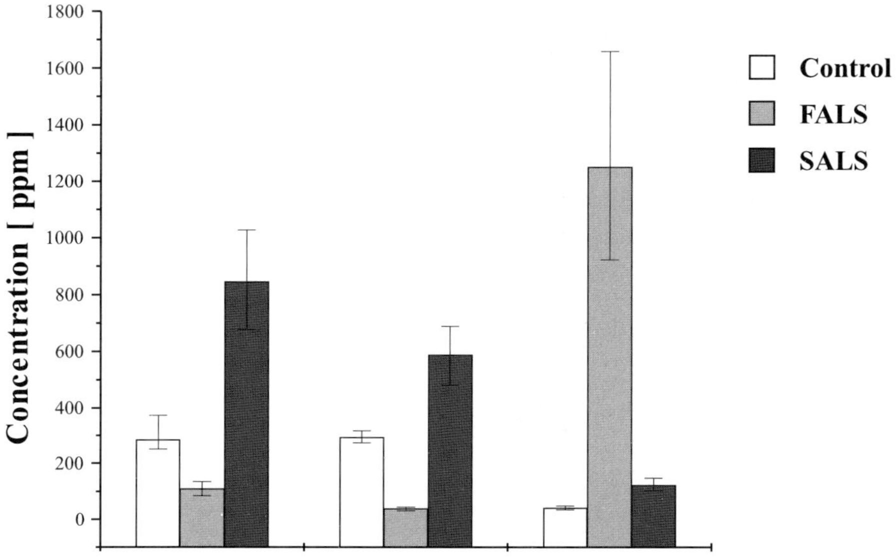

b   **Fe**   **Cu**   **Zn**

Fig. 7.7a,b. The results of quantification are shown in (**a**) table and (**b**) bar graph in ppm. The changes of the concentrations were characteristic for FALS and SALS, respectively. Interestingly, the increase of Zn was observed in both FALS and SALS cases compared to the control case

After XRF analysis, XANES analysis was performed at the beam line 39XU of SPring-8. The analysis was carried out in vacuum. The beam size was 10 µm in the diameter.

From the results of the imaging, the measurement point with the highest fluorescent intensity of Zn was selected for XANES analysis. The Zn K-edge XANES spectra obtained at the measurement point in the cells that expressed (b) GFP, (c) wild-type, (d) mutant A4T and (e) d126p are shown in Fig. 7.10. These spectra were measured in fluorescence mode. The Zn K-edge absorption spectrum was collected from 9.72 to 9.63 keV with an energy resolution of 0.5 eV and energy shifts of resolved absorption peaks of 0.25 eV can be detected. The spectrum was collected with a total signal averaging of 40 s, 50 s, 60 s/point and 60 s/point for (b) GFP, (c) wild-type, (d) mutant A4T and (e) d126p respectively. Each spectrum represents the ratio I_f/I_0

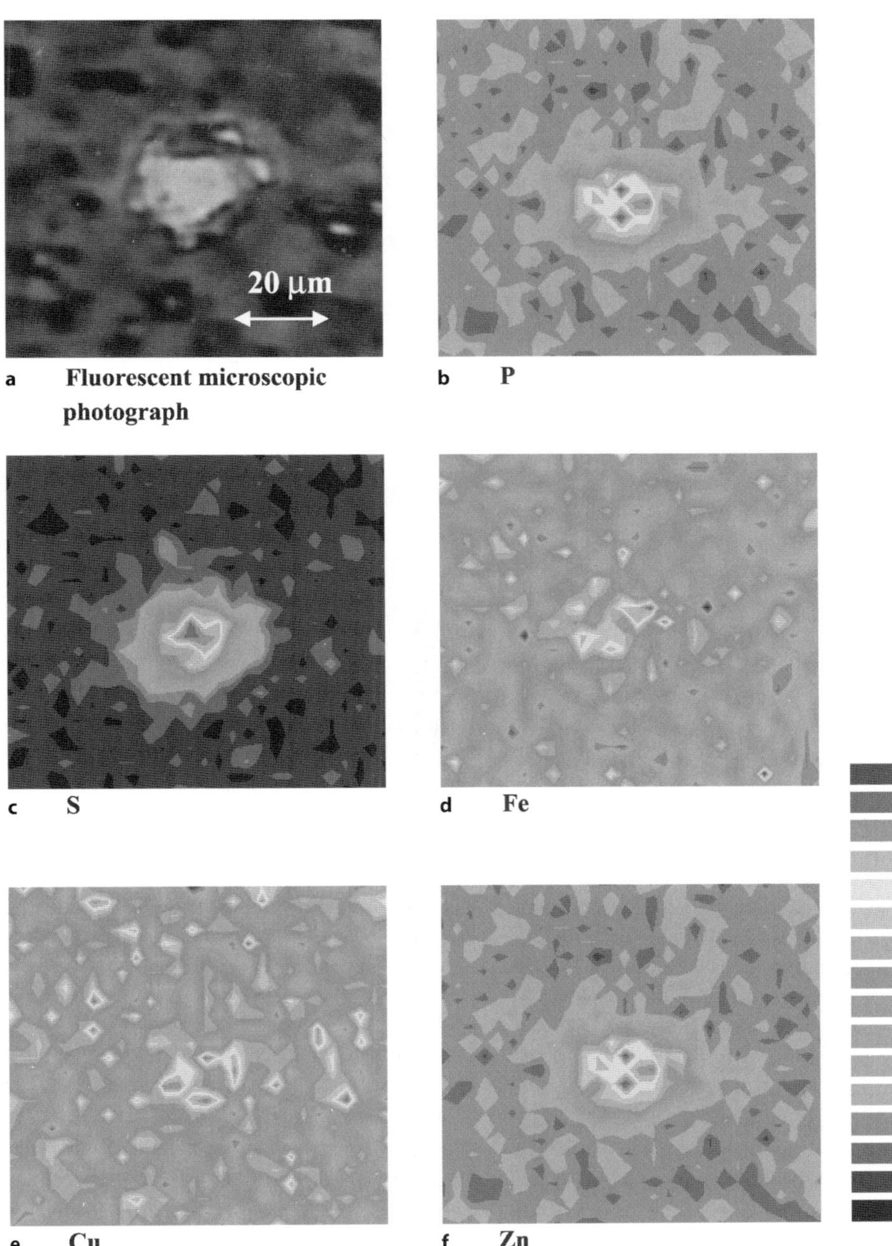

Fig. 7.8a–f. The elemental distributions of (**b**) P, (**c**) S, (**d**) Fe, (**e**) Cu and (**f**) Zn in the cells expressing wild-type SOD1. Measurement area was $81 \times 72\,\mu m^2$ and the measurement time was $8\,s/point$. The intracellular Fe and Cu contents could not be observed due to the low concentrations

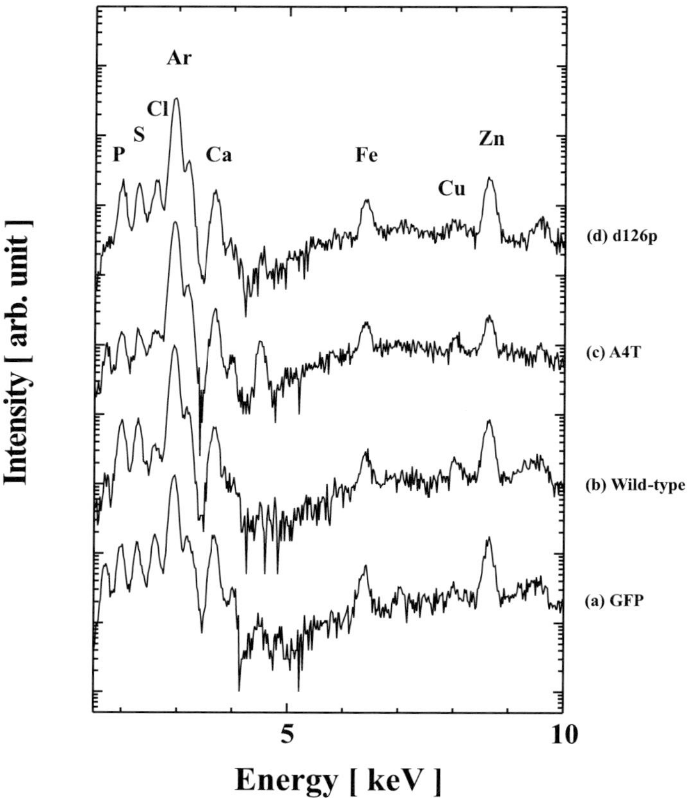

Fig. 7.9. The typical XRF spectra that were obtained in the cells expressing (a) GFP, (b) wild-type, (c) mutant A4T and (d) d126p SOD1. The measurement time was 200 s at each point. The incident beam energy was 14.3 keV. Each spectrum is normalized by the incident X-ray intensity. These cells contained P, S, Cl, Ca, Fe and Zn

(I_f = fluorescence counts, I_0 = photon incident flux measured by ionization chambers) as function of photon energy. The spectra obtained from reference samples of (a) ZnO and (f) Zn are also shown. The spectra of reference samples were collected in transmission mode and each spectrum represents the value of $-\exp(I/I_0)$ (I = transparence counts, I_0 = photon incident flux) as function of photon energy. The measurement time for these reference samples was 3 s/point. All spectra are normalized with respect to both maximum and minimum intensities.

Table 7.1. The quantification results obtained by processing XRF spectra with the computer program that was introduced in Chap. 2. The concentrations of P, S, Cl, Fe and Zn contents in the cells expressing (a) GFP, (b) wild-type, (c) mutant A4T and (d) d126p SOD1 were quantified and shown in ppm

	P	S	Cl
GFP	$15{,}038.0 \pm 7{,}729.7$	$7{,}897.5 \pm 3{,}412.9$	$6{,}101.8 \pm 2{,}732.5$
wild	$11{,}249.4 \pm 4{,}293.1$	$5{,}669.3 \pm 1{,}657.7$	$1{,}573.2 \pm\ \ \ 286.9$
d126p1	$3{,}508.5 \pm 1{,}987.7$	$1{,}456.0 \pm\ \ \ 951.3$	$995.8 \pm\ \ \ 272.8$
A4T2	$6{,}979.7 \pm 2{,}500.7$	$3{,}239.7 \pm 1{,}246.2$	$2{,}344.1 \pm\ \ \ 570.6$

	Fe	Zn
GFP	172.9 ± 81.8	214.9 ± 85.5
wild	97.1 ± 32.2	135.8 ± 33.3
d126p1	118.9 ± 27.4	63.8 ± 20.2
A4T2	114.7 ± 19.9	111.3 ± 34.9

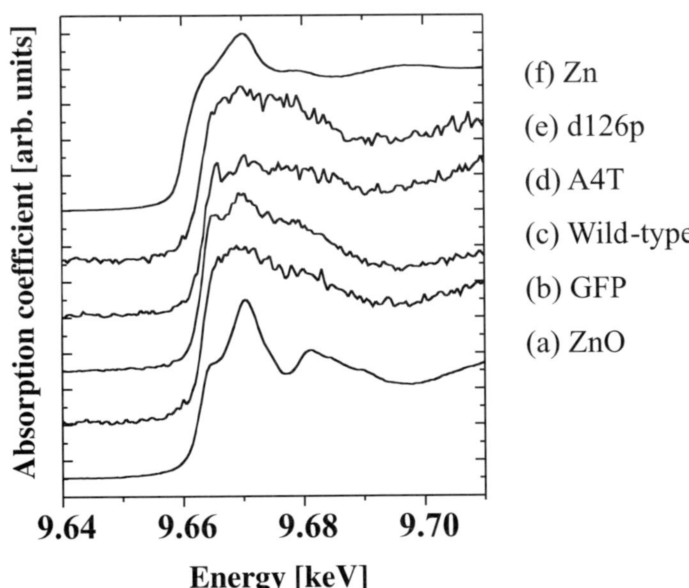

Fig. 7.10. The Zn K-edge XANES spectra obtained at the measurement point in the cells expressing (b) GFP, (c) wild-type, (d) mutant A4T and (e) d126p SOD1. The spectra obtained from reference samples of (a) ZnO and (f) Zn are also shown. The four spectra obtained in the cells are almost identical

7.2.4 Discussion

7.2.4.1 Anterior Horn Tissues from FALS and SALS Cases

The distributions of cellular elements in the control shown in Fig. 7.3 revealed that the motor neuron contained more Fe, Cu and Zn than the surrounding neuropil tissue. The red parts seen in the center of the images correspond to the motor neurons. Figure 7.5 also shows that these transition metal elements were contained in the neuron of the SALS case. In Fig. 7.4, however, the distribution of Cu in the FALS case was not detected because of its low concentration while that of Zn is clear.

The XRF spectra in Fig. 7.6 show the accumulation of Ca in the samples obtained from the FALS and SALS cases. The Ca peaks of these spectra are higher than that of the control case. It is generally considered that the mechanism of cytotoxic injury of neurons involves excessive entry of extracellular Ca through the receptors. The increased glutamate levels in cerebrospinal fluid that increases free Ca through the direct activation of Ca-permeable receptors or voltage-gated Ca channels are found in patients with SALS [18]. The finding of Ca accumulation is supportive for this mechanism and the evidences that Ca had accumulated in motor neurons through cerebrospinal fluid.

The peak of As is also clearly detected in the spectrum from the SALS case in Fig. 7.6. As mentioned in the previous paragraph, As has the direct toxicity to cells. It is considered that the absorption of As into motor neuron is deeply related to the apoptosis in the final stage of pathology as well as Ca.

The quantification results of the Fe, Cu and Zn elements shown in Fig. 7.7 revealed interesting differences among the samples from the control, FALS and SALS cases. The sample from the FALS contained Fe and Cu with 110 ppm and 39 ppm, and these are less than the control (284 ppm and 293 ppm, respectively). On the other hand, the concentration of Zn (1,250 ppm) was much higher than the control (41 ppm). Concerning sporadic ALS, the concentrations of these elements had increased to 2 to 3 times higher than the control (845 ppm, 588 ppm and 124 ppm, respectively).

These results revealed that the concentrations of metal elements in motor neurons had changed due to the progress of ALS. And the patterns of the concentration change were characteristic in familial and sporadic ALS, respectively. The changes of concentrations of Fe and Cu in patients with familial and sporadic ALS are contrary to each other. This fact may be a significant difference between the mechanisms of familial and sporadic ALS.

The decrease of Cu found in the FALS case indicates the possibility of the leak of Cu from the mutated Cu/Zn SOD, which was confirmed by Ogawa et al. [19]. This experimental result is valuable because the Cu release has actually never been identified in the tissue of FALS patients.

The most interesting finding is that the concentrations of Zn had increased in both familial and sporadic cases. According to the reports by Beckman

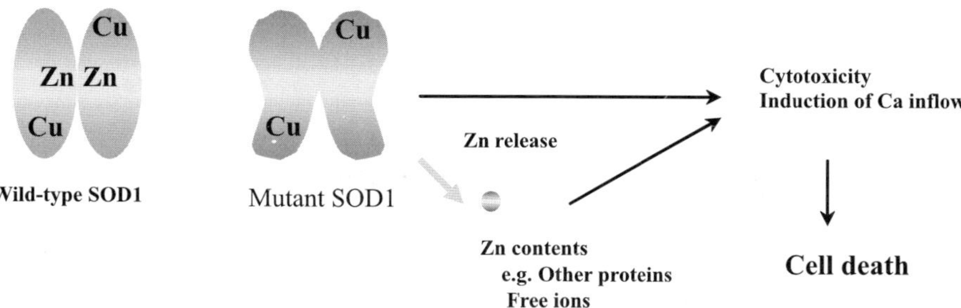

Fig. 7.11. Schematic representing the potential mechanism underlying FALS. This model was suggested through the quantitative XRF analysis in this work. Zn-deficient SOD has the cytotoxicity by itself. Furthermore, the substances generated by the release of Zn from mutant SOD such as metalloenzyme or free ions induce the new toxicity and enhance the absorption of Ca into the motor neurons, which leads to neuronal cell death

et al., mutated SODs are easy to leak Zn from their active centers and the loss of Zn promotes the cytotoxicity of mutated SOD. If the deterioration of Zn is true, the motor neurons in patients with ALS are considered to have contained Zn that were released from SODs, which exists as free ions or bound with other proteins. It is possible that these ions or proteins induce or form new cytotoxic substances. Furthermore, the increase of Zn was conspicuous in FALS rather than SALS. It is considered that the mutation of SOD and the misfold of enzymes lead to the marked accumulation of Zn.

From the results obtained in the present study, the potential mechanism that underlies FALS can be modeled as follow. Primary the mutated SODs release Zn from their active centers, and released Zn accumulated in the motor neurons as free ions and/or metalloproteins. Zn-deficient SOD has the cytotoxicity by itself through diminishing superoxide scavenging and increasing tyrosine nitration [9]. Furthermore, the substances generated by the release of Zn from mutant SOD such as metalloenzyme or free ions induce the new toxicity and enhance the absorption of Ca into the motor neurons, which leads to neuronal cell death. This mechanism is schematically presented in Fig. 7.11.

To confirm this mechanism, it is important to analyze what kind of Zn-related substances are contained in the degenerated motor neurons. If the cause of cytotoxicity of such substances is revealed, it is possible to relieve the injury of motor neurons by utilizing reagents or chelators. Recently two chelators, which remove copper from zinc-deficient SOD1 but not from normal SOD1 (replete with both Cu and Zn), have been expected to bring breakthroughs in the therapeutic field. Both chelators were proved to protect cultured motor neurons from zinc-deficient SOD1 and might be beneficial in treating human ALS [1]. By elucidating the behavior of Zn found in the FALS case, it may be possible to apply new agents for the therapy of neurodegenerative disorders.

It is also significant to clarify which stage of pathology the incursion of As and Ca was induced in. To investigate this point, more samples need to be collected in accordance with the progress of disorders.

7.2.4.2 Cultured Mouse Cells Injected with ALS DNA

The elemental imaging results shown in Fig. 7.8 shows that the cells expressing wild-type SOD1 contained P, S and Zn, while the concentrations of Fe and Cu were contained at almost the same level as outside of cells. The similar tendency in distribution of intracellular contents was also observed in the elemental imaging in the case of expression of GFP, A4T and d126p mutant SOD1 (data not shown). The concentrations of P, S, Cl, Fe and Zn are calculated from the XRF spectra obtained in the cells, from which the typical spectra are shown in Table 7.1. The quantification of Cu concentrations was impossible because the signal/noise ratio of the peaks of Cu was very low. The low concentrations of Cu even in the cells expressing wild-type SOD suggest that the quantity of SODs generated in the cells is scant.

The normalized values of the contents of Fe and Zn to the P content were compared among the cells expressing GFP, wild-type, mutant A4T and d126p because the sizes and thickness of these cells were not uniform. Phosphorous is a component of the non-diffusible proteins, and therefore it is representative of the cellular mass, as indicated by the relationship between cell mass and phosphorous concentration in the paper by Butt et al. [20].

Figure 7.12 shows that the values of Zn/P are almost identical in all samples. These Zn contents are contained in mouse Neuro2a cells or in GFP, and are not coordinated with SODs. No significant differences exist among the Zn K-edge XANES spectra obtained from the cells expressing injected DNA shown in Fig. 7.10. If the Zn affinity of wild-type SOD is deteriorated due to the mutation, it is probable that the chemical structure around Zn changes owing to the release of Zn from protein. The uniformity in the Zn K-edge XANES spectra shows that these spectra are all from the same Zn contents, which was contained in mouse Neuro2a cells and not in the wild-type or mutated SOD.

From the results of this study, Shikine et al. concluded that the purification of cells was needed in order to analyze the change in the concentrations or chemical structure of Cu and Zn using the cells injected with SOD1 gene. After the selection and centrifugation of the cells expressing SOD, it may be possible to analyze the Cu or Zn binding sites using XRF or XANES technique.

The difference in the normalized Fe content to P shown in Fig. 7.12 shows interesting characteristics. The cells expressing mutant A4T and d126p SOD1 shows the high values of Fe/P compared to those in case of expression of wild-type SOD1. Furthermore, the cells expressing d126p mutant SOD1, which showed the most severe cytoplasmic aggregates in the experiments performed by Nakano et al. [16], indicates the highest value. The normalized quantity of

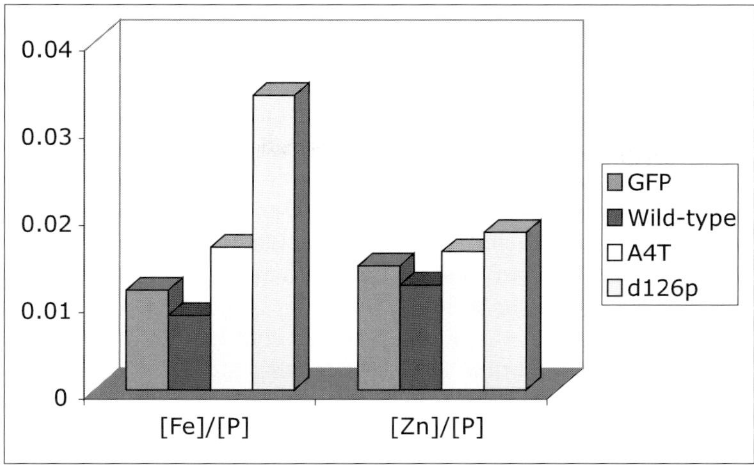

Fig. 7.12. The normalized values of Fe and Zn to phosphorus in the cells, which expressed GFP, wild-type, mutant A4T and d126p SOD1, respectively. The values of Zn/P are almost identical in all samples

Fe indicated the possibility to be the index of the cytotoxicity generated by the mutant SODs. But such increase of Fe was not observed in the analysis of the tissue obtained from the patients with FALS. Further investigation and more samples are required to reveal the difference of the metabolism in the cultured cells and in human motor neurons.

7.2.5 Conclusion

In this study, Shikine et al. investigated the correlation between neurodegenerative disorders and the intracellular elemental conditions such as distributions, concentration and chemical states using quantitative XRF analysis and XANES analysis. The trace elements in the spinal motor neurons in the anterior horn tissues of patients with ALS were investigated and compared to a control case. The non-destructive nature of SR-XRF and XANES techniques allowed observation of the same samples histologically by staining after the elemental analysis. The results obtained in the investigation are summarized below.

The motor neurons in anterior horn tissues obtained from the patients with a control, the FALS and SALS cases were analyzed using XRF technique. The results of the quantitative analysis about intracellular Fe, Cu and Zn in each sample revealed interesting changes. The sample from the FALS contained Fe and Cu with 110 ppm and 39 ppm, and these were less than the control in which these elements were contained at 284 ppm and 293 ppm respectively. The concentration of Zn was 1,250 ppm and much higher than that of the control, 41 ppm. In the sample from the SALS case, the concentrations of these elements had increased and the concentrations of Fe, Cu and Zn were 845 ppm, 588 ppm and 124 ppm respectively. These results revealed that the concentrations of intracellular Fe, Cu and had changed due to the progress of ALS, and the patterns of the concentration change were characteristic in the FALS and SALS cases, respectively. It appeared that the difference between the mechanisms had been reflected into the change of the concentrations. The increase of Zn suggested the possibility of the accumulation of Zn content that had released from SOD in the motor neurons and the generation of new cytotoxicity. The accumulation of Ca was also observed in the FALS case, and it was revealed that As had accumulated in the motor neurons in the SALS case as well as Ca.

As the model of FALS, the mouse Neuro2a cells injected with GFP, wild-type, A4T and d126p mutant SOD1 were also analyzed using XRF and XANES spectrometry.

The cells expressing mutant A4T and d126p SOD1 showed high values of Fe/P compared to those in case of expression of wild-type SOD1. The cells expressing d126p mutant SOD1 indicated the highest value. The normalized quantity of Fe may be the index of the cytotoxicity generated by the mutant SODs. But the chemical structures around the Cu or Zn binding sites in SODs could not be analyzed due to the poor quantity of expression. No significant differences could be seen in the Zn K-edge XANES spectra obtained from the cells expressing GFP, wild-type, A4T and d126p mutant SOD1. It is probable that these Zn contents were contained in mouse Neuro2a cells or in GFP, not coordinated with SODs. Shikine et al. concluded that the purification of cells is needed in order to analyze the change in the concentrations or chemical structure of Cu and Zn using the cells injected with SOD1 gene.

7.2.6 Summary

In the investigation by Shikine et al. (2002), two forms of ALS, familial ALS (FALS) and sporadic ALS (SALS) cases were analyzed and compared with a control case. The characteristic patterns in the elemental condition were observed in each case. It is probable that these characteristics reflected the mechanisms of neuronal cell death in FALS and SALS. As the significant cause of the neuronal toxicity in the pathogenesis of ALS, many researchers have reported the release of Zn from the active centers of Zn/Cu SOD. SOD controls oxygen toxicity by converting the superoxide radical to less dangerous forms. On the contrary to the prediction, the increases of Zn contents were revealed in both FALS and SALS cases. The possibility was suggested that the new cytotoxicity is generated by the accumulation of Zn contents released from SOD in the motor neurons. The accumulation of Ca was also clarified in both cases. The cytotoxicity induced by Zn contents had likely to result in the inflow of Ca causing cell death. As the model of FALS, the mouse Neuro2a cells injected with wild-type and mutant SOD1 were also analyzed using XRF and XANES spectrometry, but the chemical structures around the Cu or Zn binding sites in SODs could not be analyzed due to the poor quantity of expression.

These experimental results showed that the elemental condition of cells reflected their functions and states characteristically. The typical examples are the increase of Zn and Ca contents in FALS and SALS cases, the decrease of Cl and the increase of Zn contents in the mouse ES cells in unoriented differentiation, and the uptake of K and Mn contents in the neuronal differentiation with PA6 feeder layer.

In this study, the cultured cells injected with SOD1 gene were analyzed as the model of FALS. In order to understand the role of the intracellular specific elements, it is important to analyze the object and the modeled biological system together and to relate the results obtained from both samples. From the findings of the investigation into neurodegenerative disorders, the effect of Ca, Fe, Cu, Zn and As to the neurons seem to be significant to elucidate the mechanism of neuronal cell death. The effects of these elemental stimulations should be examined using the cultured neurons. From the differences in the concentrations of intracellular trace elements, interestingly, the new potential mechanism of the cytotoxicity generated by Zn contents was suggested in the investigation about ALS. This mechanism has not been conceived without the quantitative information in the single neuron. It is important to confirm this mechanism by removing or adding intracellular Zn contents by chelators or reagents.

7.3 Application for Investigating the Mechanisms of ALS

7.3.1 Introduction

Glutamate is the main excitatory neurotransmitter in the human nervous system. There are several subtypes of glutamate receptors, which are divided into NMDA-type and non-NMDA-type. AMPA and kainate receptors, collectively referred as non-NMDA receptors, represent two of the major classes of glutamate receptors. It has been suggested that the excessive activation of non-NMDA receptors can result in the excitotoxic death of neurons, particularly with prolonged exposure to receptor agonists [21–23]. Amyotrophic lateral sclerosis (ALS) is caused by progressive death of motor neurons. The causes of the selective vulnerability of motor neurons in ALS are still unknown. Although NMDA receptors have generally been regarded as the most likely mediator associated with the chronic neuronal injury, several observations suggest that Ca^{2+}-permeable AMPA/kainate receptors may be of greater importance to the slow neurodegenerative process seen in ALS [22,24]. Recently, Sean G. Carriedo et al. suggest that AMPA or Kainate exposures trigger substantial mitochondrial Ca^{2+} loading, and caused the mitochondrial depolarization and the reactive oxygen species (ROS) generation in motor neurons [25,26]. Furthermore, recent studies suggest that mitochondrial Ca^{2+} overload in response to Ca^{2+} entry through Ca^{2+}-permeable AMPA/kainate channels may be caused by a lack of Ca^{2+}-binding proteins in motor neurons [27,28].

In order to measure the amount of the intracellular Ca^{2+}, some useful methods have been investigated and developed for long periods of time. Radioactive calcium or the chemical fluorescent indicator, such as *Fura 2*, are often employed to measure the intracellular or subcellular calcium [29–31]. It is suggested that the selective vulnerability of motor neurons is caused by Ca^{2+} loading in motor neurons through AMPA/Kainate receptor channels in vitro [22,26,32]. However, there are few studies about the in vivo measurement of Ca^{2+} loading in motor neurons through AMPA/Kainate receptor channels.

Kitamura et al. (2000) conducted a study of aimed at carrying out in vivo measurement of the influx of Ca^{2+} into motor neurons through AMPA/Kainate receptor channels in the generation of excitotoxic injury induced by injecting kainic acid into the peritoneum of a Wistar rat. In this study, the in vivo measurement of the distributions of calcium and other elements in neurons was realized by using SR-XRF spectrometry. In order to measure the intracellular calcium in vivo, the freeze dried sample of the spinal cord of a rat was employed.

7.3.2 Material and Methods

In this study, Kitamura et al. used male Wistar rats, weighing about 300 g. Kainic acid dissolved in artificial cerebrospinal fluid was administered into the intrathecal of the rat via an inserted tube by mechanical pump at a rate of 1 nmol/h for 2 months. After the spinal cords were removed from the rats, they were frozen rapidly by using liquid nitrogen and thin sections of 10 μm thickness were cut in a cryostat. The thin sections were mounted on PET films and dried for SR-XRF measurement. As the control case, only the artificial cerebrospinal fluid was administered into the intrathecal of the Wistar rat under the same conditions.

In order to compare with the chronic effect of the kainic acid, the acute effect of the kainic acid was investigated. Kainic acid (10 mg) dissolved in artificial cerebrospinal fluid was administered into the intrathecal of the rat. After an hour, the brain was removed from the rat and frozen rapidly using liquid nitrogen. Sections of 10 μm thickness were cut, dried and mounted on the PET films for the XRF measurement. The hippocampal CA1 pyramidal cells were measured by XRF analysis.

SR-XRF analysis was performed at the Photon Factory BL4A of the High Energy Accelerator Research Organization, Tsukuba, Japan. The excitation beam energy was 14.2 keV, and beam size was 10 μm. All experiments were done in the air.

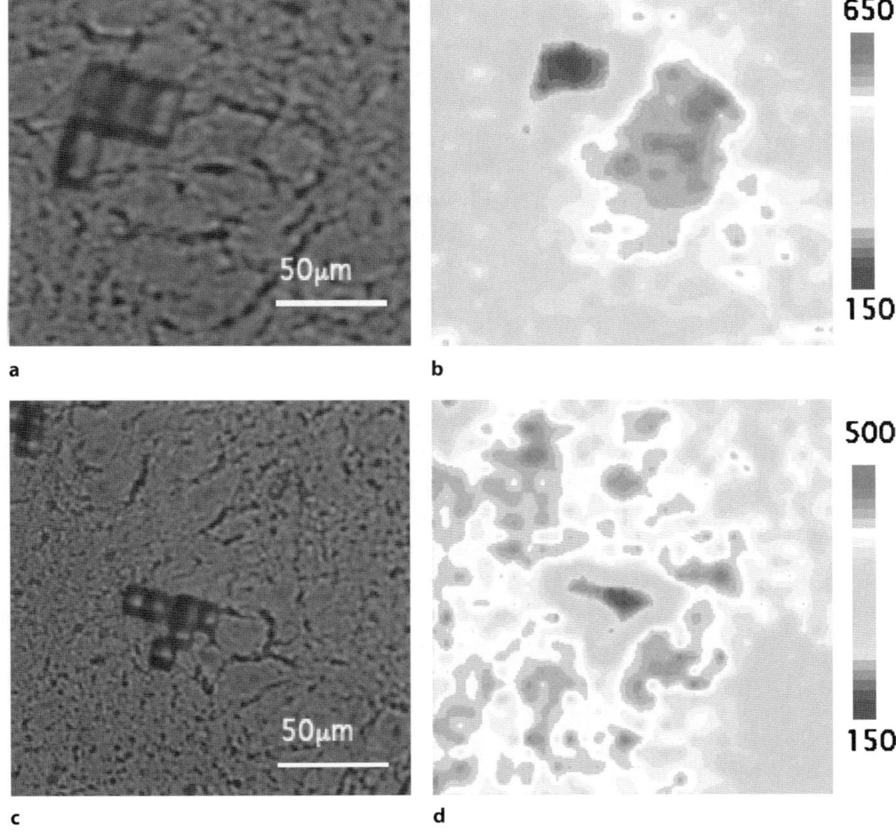

Fig. 7.13a–d. The small burned points were plotted near the aimed neurons. When the elemental distribution patterns of the neurons with the burned marks are obtained by XRF analysis, the concentration of K in the burned area is lower than the other area

7.3.3 Experimental Results

All measured samples were marked by a laser beam in order to distinguish the target cells from the other ones. The small marked points were plotted near the target neurons. When the elemental distribution patterns of the neurons with the marks are obtained by SR-XRF analysis, the concentration of potassium at the marked point is lower than the other area. If the marked points are plotted to draw the distinctive pattern as shown in Fig. 7.13a,c, the concentrations of potassium within these area are lost as shown in Fig. 7.13b,d, respectively. With the help of these marks, the target cells can be distinguished perfectly.

Acute Effect of Kainic Acid

In order to observe the inflow and accumulation of calcium within the hippocampal CA1 pyramidal cells through NMDA and voltage-sensitive Ca^{2+} channels (VSCC), kainic acid was injected into the rat rapidly as described above. The optical light microscopic photograph of the hippocampal CA1 pyramidal cells is shown in Fig. 7.14. The elemental distributions in the red square part shown in Fig. 7.14 were measured by employing SR-XRF imaging technique. The elemental distribution patterns of Ca, Fe and Zn are shown in Fig. 7.15, prepared as previously. The scanning area was $30 \times 30\,\mu m^2$, in $1\,\mu m$ pixel resolution. It is considered that calcium, iron and zinc may be flown into the CA1 pyramidal cell through NMDA channels and VSCC owing to the rapid and strong stimulation induced by injecting kainic acid. The XRF spectrum in Fig. 7.16 was obtained at the point with the highest density of Ca within the area shown in Fig. 7.15. In addition, Fig. 7.17 shows a spectrum from a point selected from outside the accumulation point of Ca. The value of the density of Ca calculated from the spectrum shown in Fig. 7.16 is about twofold higher than that shown in Fig. 7.17 and 1.8 times higher than the value of the control cell. The fluorescent X-ray intensities of Zn and Fe within the control cell were very low. It is probable that the inflow of Zn into this pyramidal cell may be accompanied by that of Ca because of the fact that Zn has the same valence state as Ca. If the excessive accumulation of Ca becomes an irreversible state, the neurodegeneration will start. Furthermore, it is necessary to clarify the reason for the accumulation of Fe within the cell, which may be accompanied by that of Ca. It is necessary to perform further investigations into the influence of the accumulation of Zn and Fe within the pyramidal cell.

Chronic Effect of Kainic Acid

The control motor neuron in the spinal cord was stained by 0.1% toluidine blue solution and is shown in Fig. 7.18. The XRF spectrum measured at a point in the control motor neuron is shown in Fig. 7.19. In order to measure Ca^{2+} loading within the spinal motor neurons through AMPA/Kainate receptor channels, kainic acid was injected into the rat slowly as described above. The motor neurons and other glial cells treated by the slow injection of kainic acid were stained by 0.1% toluidine blue solution and is shown in Fig. 7.20. The red square areas named No.1, No.2 and No.3 include the motor neurons and glial cells, and these areas were measured by SR-XRF imaging technique.

   The elemental distribution patterns of Ca, Fe and Zn were obtained from No.1 area and are shown in Fig. 7.21. This image is a matrix of 40×40 pixels of $1\,\mu m$ resolution. The overlapping areas of the elements are expressed by their superimposed colors. Ca, Fe and Zn are localized in different parts of the cell. It is plausible that Fe and Zn may be flown into the cell together with Ca through AMPA/kainate receptor channel. In general, Ca is said to be

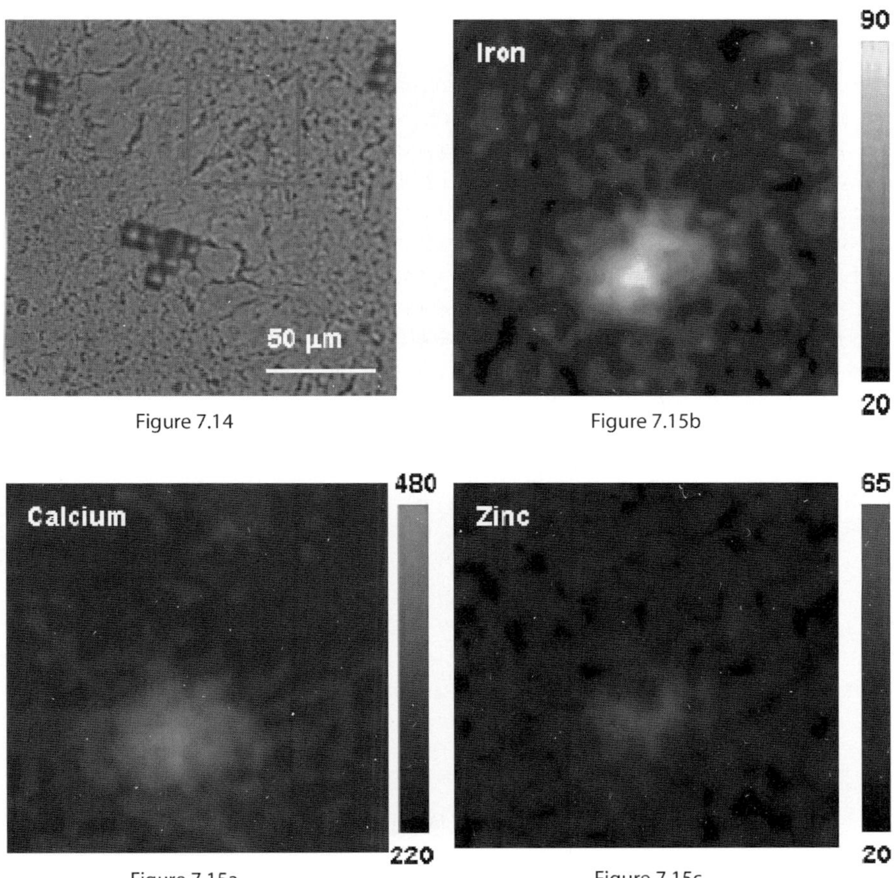

Fig. 7.14 and 7.15a–c. The optical light microscopic photograph of the hippocampal CA1 pyramidal cells was shown in (**7.14**). The elemental distribution patterns of Ca, Fe and Zn in this area are shown in (**7.15a–c**). The scanning area was $30 \times 30 \, \mu m$ and these images are matrices of 30×30 pixels of $1 \, \mu m$ resolution

impermeable through the AMPA/Kainate receptor channel. Yet, it has been suggested that there is Ca^{2+}-permeable AMPA/Kainate receptor channel as the subtype of AMPA/Kainate receptor channel. It is probable that Ca and Zn may have flown into the cell through Ca^{2+}-permeable AMAPA/Kainate receptor channel.

The XRF spectra were obtained at the points with the highest densities of Ca and Zn and these are shown in Figs. 7.22 and 7.23, respectively. The density of Ca within the cell shown in Fig. 7.21 is 2.6 times higher than that within the control cell and 3.0 times higher than that at a point outside the cell. Furthermore, the accumulation of Zn is accompanied by the inflow of calcium. It is not clear why the elemental distribution patterns of Ca, Fe and Zn display the separation and localization.

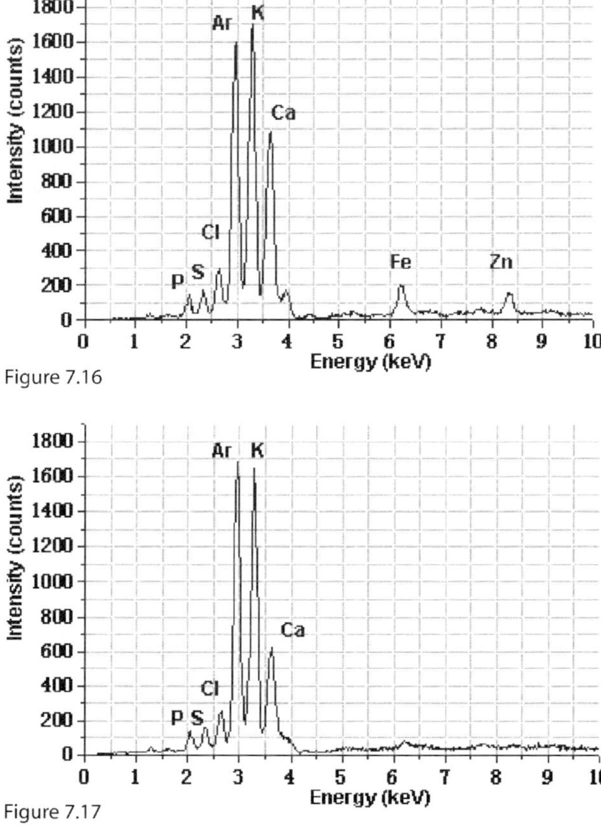

Fig. 7.16 and 7.17. The XRF spectrum was not obtained at the point with the highest density of Ca within the area shown in Fig. 7.15 and it is shown in (**7.16**). On the other hand, a point was selected from outside the accumulation area of Ca, and the XRF spectrum was obtained at this point. This spectrum is shown in (**7.17**). The value of the density of Ca calculated from the spectrum shown in (**7.16**) is about 2 times as high as that shown in (**7.17**) and 1.8 times as high as the value in the control cell

The elemental distribution patterns shown in Fig. 7.24 of Ca and K were obtained in the No.2 area shown in Fig. 7.20. The scanning area was 60 × 60 μm^2 and this image is a matrix of 30 × 30 pixels of 2 μm resolution. It appears that the distribution of K spread in overall of the motor neuron. It is observed that Ca is localized in the motor neuron. The two XRF spectra shown in Figs. 7.25 and 7.26 were obtained at the point with the highest density of Ca and at a point next to it inside the motor neuron, respectively. Integrating the intensity value of Ca, the highest value is 2.5 times higher than that in the control motor neuron, and 3.2 times higher than that in the point next to it.

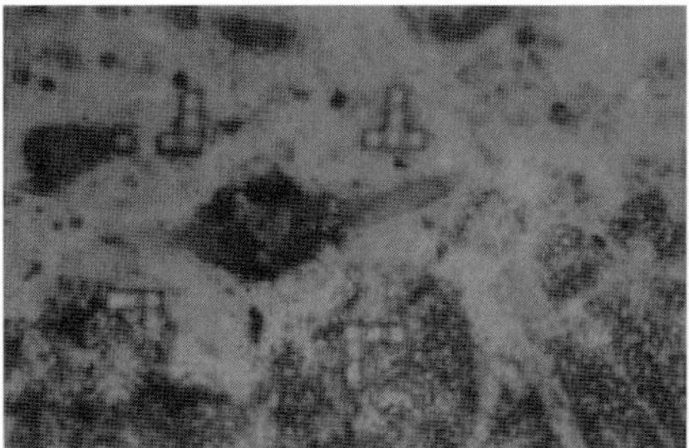

Figure 7.18

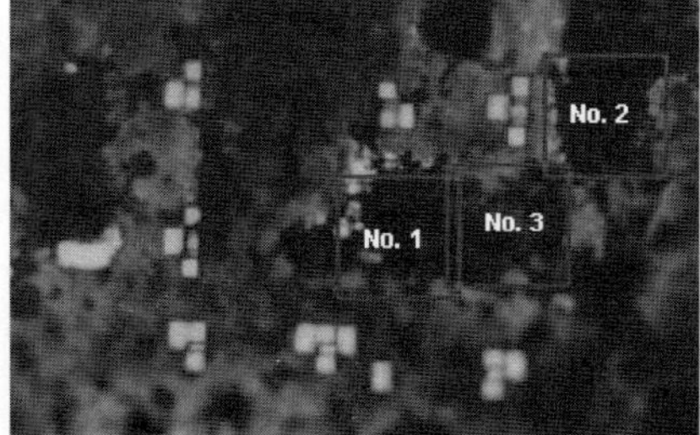

Figure 7.20

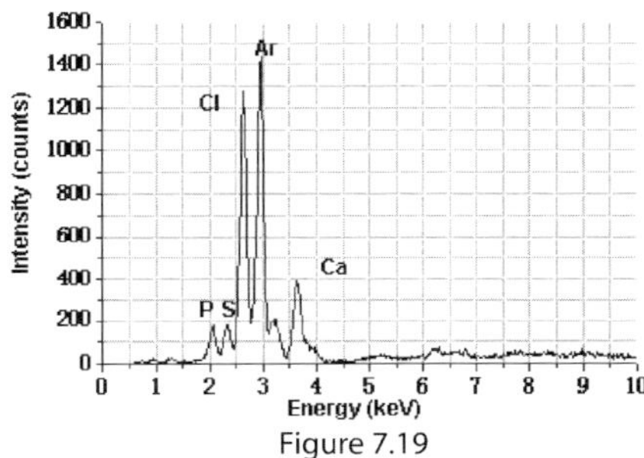

Figure 7.19

Fig. 7.18, 7.19 and 7.20. The control motor neuron in the spinal cord was stained by 0.1% toluidine blue solution and shown in (**7.18**). The XRF spectrum measured at a point in the control motor neuron is shown in (**7.19**). The motor neurons and other glial cells treated by the slow injection of kainic acid were stained by 0.1% toluidine blue solution and shown in (**7.20**). No.1, 2 and 3 area include the motor neurons and glial cells, and these areas were measured by SR-XRF imaging analysis

The elemental distribution patterns displayed in Fig. 7.27 of Ca and K were obtained in the No.3 area shown in Fig. 7.20. The scanning area was $60 \times 60\,\mu m^2$ and the matrix is of 30×30 pixels of $2\,\mu m$ resolution. It is observed that Ca is localized in the motor neuron. The two XRF spectra shown in Fig. 7.28 and Fig. 7.29 were obtained at the point with the highest density of Ca and at a point next to it inside the motor neuron, respectively. In this case, the highest value of the density of Ca in the motor neuron shown in Fig. 7.27 is 1.7 times higher than that in the control motor neuron, and 2.0 times higher than that for the point next to it in the motor neuron.

7.3.4 Discussion

Kitamura et al. observed that the overload of calcium within the CA1 pyramidal cells was induced by the rapid inflow of calcium through NMDA receptor channels and voltage sensitive calcium channels. Furthermore, zinc and iron were detected in the cell, too. However, although the excitation of the neurons

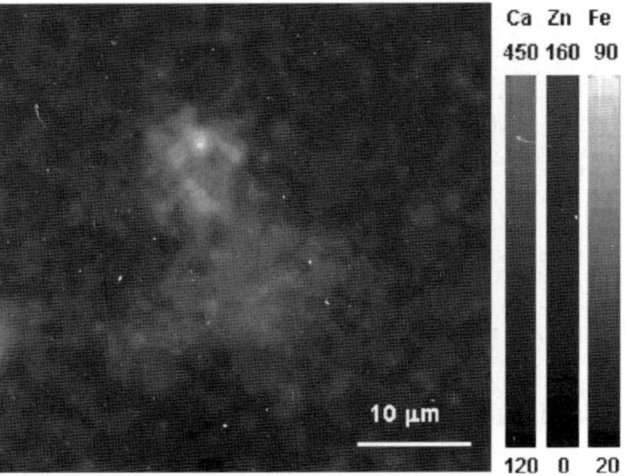

Figure 7.21

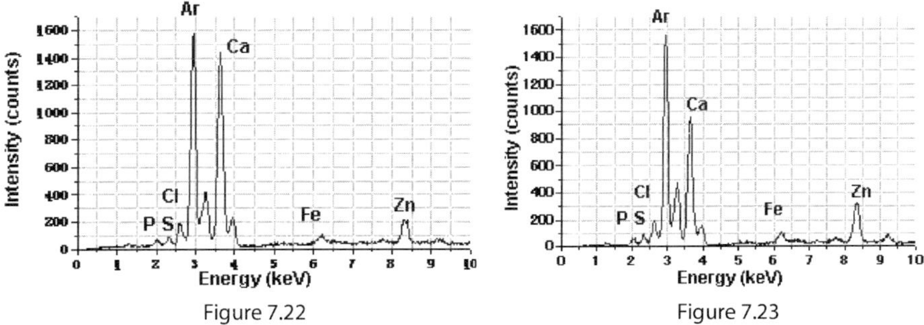

Figure 7.22   Figure 7.23

Fig. 7.21, 7.22 and 7.23. The elemental distribution pattern of Ca, Fe and Zn were obtained from the No.1 area shown in Fig. 7.20, and shown in (**7.21**). The scanning area was $40 \times 40\,\mu$m. This image was matrices of 40×40 pixels of $1\,\mu$m resolution. The overlapping area of the elements are expressed by their superimposed colors. The XRF spectra were obtained at the points with the highest densities of Ca and Zn, and are shown in (**7.22**) and (**7.23**), respectively

was extremely strong, not all of the pyramidal cells have the localized distribution patterns of calcium. It is necessary to perform further investigations into the reason for this phenomenon.

For the chronic effect of the slow injection of kainic acid into the rat, the overload of calcium within the spinal motor neurons was observed. The inflow of calcium into the cell may be induced through Ca^{2+}-permeable AMPA/Kainate receptor channels. There were no differences between the values of calcium densities within the pyramidal cells and spinal motor neurons after the rapid and chronic excitation of the neurons by injecting kainic acid. It can be concluded that the irreversible overloads of calcium within the spinal motor neurons through Ca^{2+}-permeable AMPA/kainate receptor channels had occurred before the neuronal degenerations.

It is assumed that the overload of calcium is observed in the mitochondria and the consequent production of injurious oxygen radicals may have

Figure 7.24

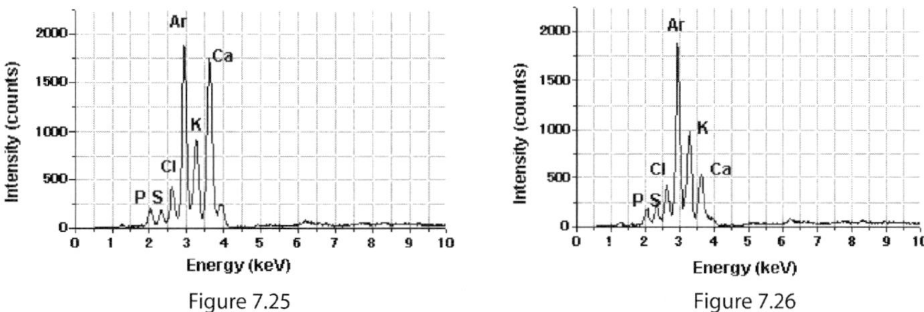

Figure 7.25   Figure 7.26

Fig. 7.24, 7.25 and 7.26. The elemental distribution pattern of Ca and K were obtained from No.2 area shown in Fig. 7.20, and shown in (**7.24**). The scanning area was $60 \times 60\,\mu$m. This image was matrices of 31×31 pixels of $2\,\mu$m resolution. The overlapping area of the elements are expressed by their superimposed colors. The two XRF spectra were obtained at the point with the highest density of Ca and at the point inside the motor neuron beside the Ca accumulation area, respectively. The former spectrum is shown in (**7.25**) and the latter is shown in (**7.26**)

occurred [25]. Therefore, it is probable that the accumulation of calcium in the spinal motor neurons shown in this experiment may have occurred in the mitochondria. It is indicated that the exitotoxic injury of the spinal motor neurons can be caused by the weak and continued excitation of the neurons because of the irreversible overload of calcium into the motor neurons through Ca^{2+}-permeable AMPA/Kainate receptor channels. ALS causes the selective vulnerability of the spinal motor neurons. The reason for this selective degeneration of the motor neurons is still unknown, but it is suggested that the selective vulnerability of motor neurons is caused by the expression of highly Ca^{2+}-permeable AMPA receptors [32]. The overload of calcium in the spinal motor neuron shown in this experiment is considered to be relevant to the neuronal degenerative mechanisms of ALS.

In vivo measurement of the overload of calcium in the spinal motor neurons by the injection of kainic acid into the rat was made possible by using the microbeam from synchrotron radiation source.

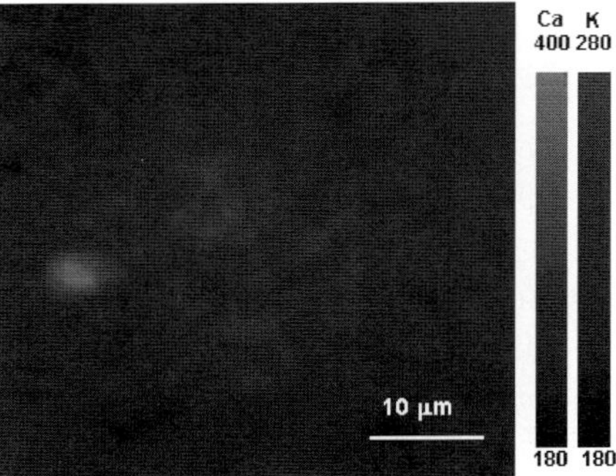

Figure 7.27

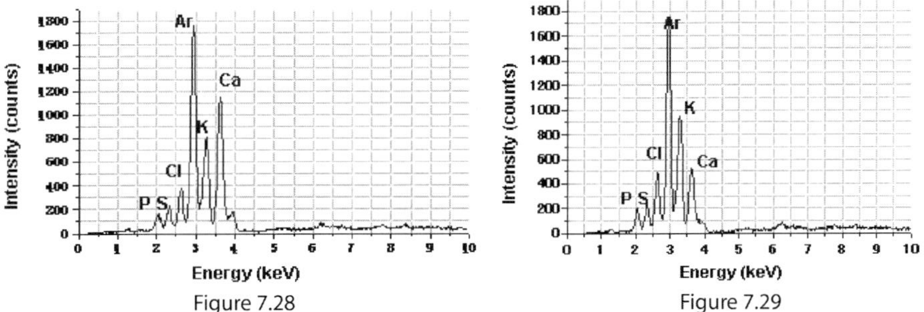

Figure 7.28   Figure 7.29

Fig. 7.27, 7.28 and 7.29. The elemental distribution patterns of Ca and K were obtained from No.3 area shown in Fig. 7.20, and shown in (**7.27**). The scanning area was $60 \times 60\,\mu m$. This image was matrices of 31×31 pixels of $2\,\mu m$ resolution. The overlapping area of the elements are expressed by their superimposed colors. The two XRF spectra were obtained at the point with the highest density of Ca and at the point inside the motor neuron beside the Ca accumulation area, respectively. The former spectrum is shown in (**7.28**) and the latter is shown in (**7.29**)

7.4 Quantitative Analysis of Zinc, Copper and Iron in Alzheimer's Disease

7.4.1 Introduction

Alzheimer's disease (AD) is a progressive neurodegenerative disorder, which affects memory, thinking, behavior and emotion. The pathogenesis of the disease is still not clear, but recent molecular basis studies suggested that the deposition of amyloid-β (Aβ) peptide, oxidative stress and mutation of presenilin contribute to the neuronal degenerations in AD. Metallic elements such as zinc, copper and iron are proposed as being closely related to the mechanism of degeneration and cell death, associated with these molecules.

Several hypotheses of the metal-associated mechanisms of the disease have been proposed:

- *Aβ and metallic elements*
  The deposition of Aβ, which is associated with neuronal death and oxidative stress, is considered to be responsible for the primary pathogenesis of AD [33]. Aβ possesses selective affinity Cu^{2+} and Zn^{2+} binding site [34–36], and copper and zinc ions are closely related to induction of the dissoluble deposition of Aβ [37,38]. Moreover, metals binding Aβ may play an important role to mediate the generation of reactive oxygen species. Aβ is redox active, and produces hydrogen peroxide through metal ion depletion in vitro [39,40]. These studies propose that metal-mediated redox activity of Aβ may play an important role in the pathogenesis of AD [41], but it is not clear how metals bind Aβ in vivo.
- *Oxidative stress and metallic elements*
  Reactive oxygen species (ROS) and the resulting oxidative stress also play an important role in the pathogenesis of AD. Transition metals such as iron are considered to be one of the important sources of radicals in the disease. In AD, alterations of iron metabolism are observed, including increased levels of free iron as well as altered levels of iron transport and storage proteins [42–45]. Furthermore, redox-active iron is associated with the pathological lesions [44], and it is suggested that iron accumulation is a source of redox-generated free radicals in AD.
- *Presenilin mutations and Calcium homeostasis*
  Recent findings suggest links among deposition of Aβ, oxidative stress, disruption of ion homeostasis, and an apoptotic form of neuron death in AD [46]. It is reported that expression of the human presenilin mutation increases their susceptibility to apoptosis induced by trophic factor withdrawal and Aβ. Increases in oxidative stress and intracellular calcium levels induced by the apoptotic stimuli were exacerbated greatly in cells expressing the PS-1 mutation [46,47].

These researches show that the role of metals in neuronal degeneration in AD is significant *in vitro*. Therefore, it is of interest to know the behavior of metallic elements in the tissues of AD brains.

In a study, Fujisawa et al. (2002) investigated the distribution and chemical state of metals in the post-mortem tissues of AD, aimed at analyzing metallic elements contained in the tissues directly. They applied synchrotron radiation X-ray fluorescence analysis and X-ray absorption near-edge structure analysis, to the unstained section. The XANES analysis was used to obtain the chemical state of iron contained in the tissues of AD.

7.4.2 Experimental Procedures and Results

The samples consisted of post-mortem brain tissues of AD patients (Table 7.2). Tissue blocks stored snap-frozen at $-80\,°C$ were dissected from tem-

Table 7.2. Summary of the cases, all of which were obtained at autopsy

Case	Gender	Age	Diagonosis	Region to study	
1	F	87	SDAT + CVD	Temporal lobes	8904
2	F	82	SDAT + CVD	Temporal lobes	9001
3	F	64	AD	Hippocampus	9503
4	F	62	AD	Hippocampus	9905

SDAT: Senile dementia of Alzheimer Type
CVD: Cerebrovascular disease

poral lobes of the right hemisphere cortex and fixed in 2% glutaraldehyde (pH 7.4) for 2 h. Sections of 5 µm thickness were made from the fixed blocks and dried at room temperature, followed by mounting on 25 µm thick PET films for X-ray analyses.

X-ray fluorescence (XRF) analyses were performed at beam line 4A of Photon Factory, High Energy Accelerator Research Organization (KEK), using a $7 \times 5\,\mu m^2$ incident beam. Measurements were performed in air. The X-ray absorption fine structure analyses were performed in vacuum at beam line 39XU of SPring-8, using a 10 µm diameter beam. Fe K-edge XANES analyses were performed in the energy range of 7.100 to 7.160 keV at 0.5 eV intervals. The data were measured in fluorescence mode for biological specimens and in transmission mode for the reference samples. Incident and transmitted photon flux was monitored with an air-filled ion chamber. Fe K-edge fluorescent X-rays were also collected by a solid state detector (SSD).

Figure 7.30 shows the optical microscopic photograph of a section stained for ubiquitin. In this figure, the section was ubiquitin immuno-reactive. Ubiquitin immunoreactivity is used regularly in the identification of pathological lesions, such as Lewy bodies and neurofibrillary tangles (NFT), that are associated with neurological disorders [48]. Abnormal levels of ubiquitin in the brain of AD patients have been proposed as markers of the disease [49]. The XRF analyses were performed to the tissue in the parallel unstained section. First, the tissue area where the ubiquitin deposition was remarkably observed in the stained section was analyzed. XRF images (40×40 pixels of 1 µm size) of Fe, Zn, Ca and S, and an optical microscopic photograph in the tissue area are shown in Fig. 7.31. The ranges of fluorescent X-ray intensity are shown in the color scales (arbitrary unit). Features of a cell can be observed in the scanning area. Fe and Zn concentrations are detected in the cell, and the distributions of Fe and Zn are different. A typical X-ray spectrum in the cell is shown in Fig. 7.31f. The cell contains many metallic elements, such as Ti (2,904 ppm), Fe (1,241 ppm), Cu (586 ppm) and Zn (1,227 ppm).

Figure 7.32 shows the results for another area in the tissue. The images have 40×50 pixels of 1 µm size. A typical X-ray spectrum from a cell in this area is shown in Fig. 7.32f. The spectra show that Fe, Cu and Zn are

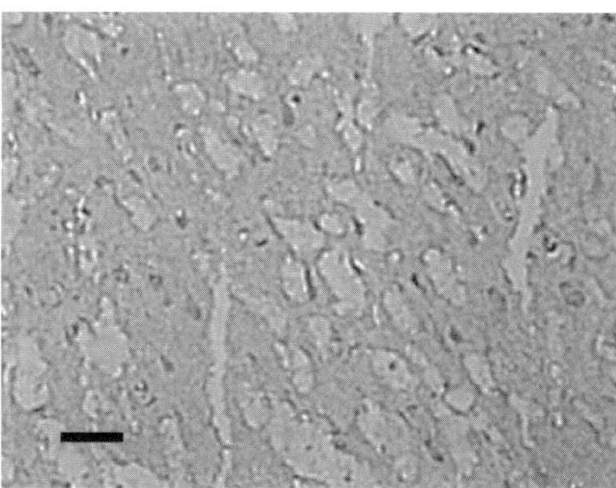

Fig. 7.30. The optical microscopic photograph of the section stained for ubiquitin. The scale bar is 50 μm

contained in the cell, but the yields of fluorescent X-rays are not so large. Quantification results of the spectrum gave the density of Fe, Cu and Zn as 352, 88 and 143 ppm, respectively. These and those from a random selection of neurons from a section of patient A are shown in Table 7.3, where the increase of the densities of Fe, Cu and Zn in the ubiquitin deposited tissues can be observed.

The Fe and Ca XRF images (60×30 pixels of 1 μm size) from a specimen obtained from the hippocampus of AD patient (case 3) are shown in Fig. 7.33. Features of a cell in the Fe scanning area can be seen. The typical X-ray spectrum in this cell is shown in Fig. 7.33c. Fe concentration is high in the cell (about 13000 ppm), but Cu and Zn are not detected.

Fujisawa et al. (2002) also analyzed the chemical states of iron contained in the temporal lobes tissues of the patient (case 1), using the Fe K-edge XANES analyses. Figure 7.34 shows XANES spectrum at a cell in the tissues and those of reference samples (FeO and Fe_2O_3). The ordinate and abscissa represent the absorption coefficient and incident X-ray energy, respectively. The spectra were normalized by the absorption jump. Absorption jump was defined as the difference between the highest and the lowest point in each spectrum. The absorption edge is defined at the half-height of the absorption jump. From the spectra of Fig. 7.34 the chemical shifts of Fe contained in the cell of the AD case can be derived to be 3.9 eV from that of Fe^{2+} in FeO, while the chemical shift of Fe^{3+} of Fe_2O_3 is about 4.3 eV. The chemical state of iron in the cell with AD has shifted toward Fe^{3+}.

7.4.3 Discussions and Summary

The toxicity of the metallic elements such as iron, copper and zinc in Alzheimer's disease (AD) has been proposed as one of the pathogenesis of

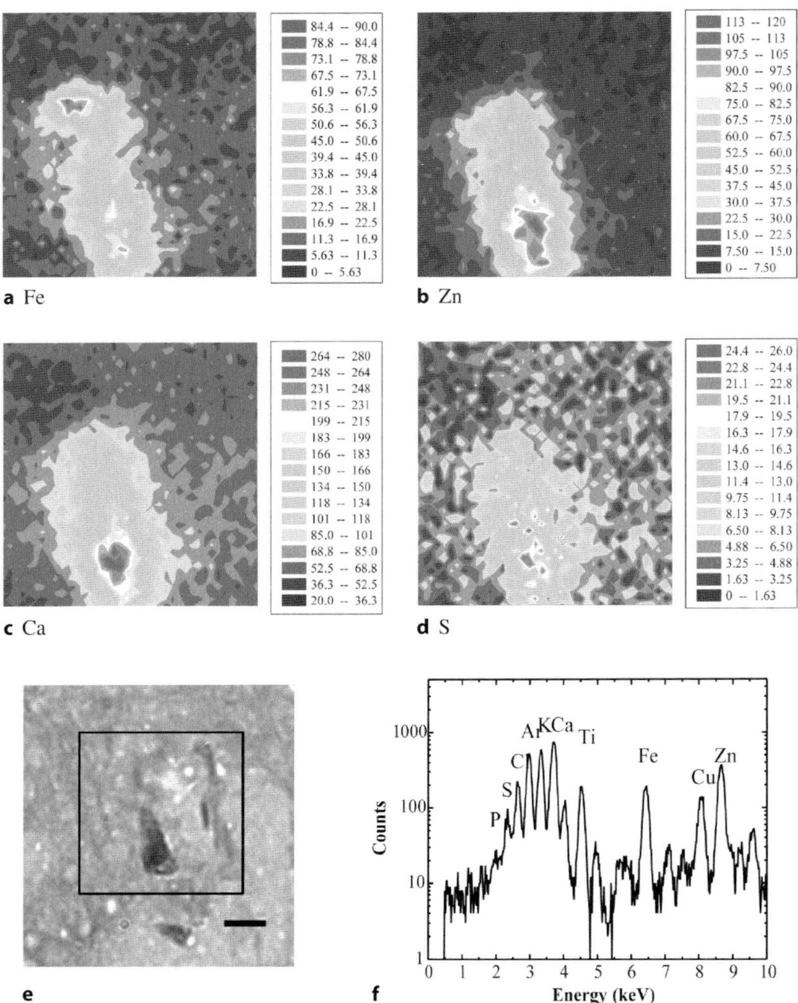

Fig. 7.31a–f. XRF images of Fe, Zn, Ca and S in the temporal lobes tissue of the AD patient (**a**)–(**d**). In this area, the tissue is ubiquitin immunoreactive. The optical microscopic photograph of the sample stained with HE (**e**). The staining was performed after X-ray analysis. The scale bar is 20 μm. Typical X-ray fluorescence spectra in the tissue (**f**). The measurement time is 200 s

the disease. In this study, Fujisawa et al. (2002) analyzed the distribution and the densities of the transition metals in the AD tissues from four patients, using X-ray SR-XRF. Significant concentrations of Zn, Cu and Fe were observed in the temporal lobes tissue with Alzheimer's disease. The quantification result shows that the average density of zinc was 342 ppm, copper was 244 ppm and iron was 1537 ppm in the AD temporal lobes of case 1. X-ray absorption near-edge structure (XANES) analysis was also applied in

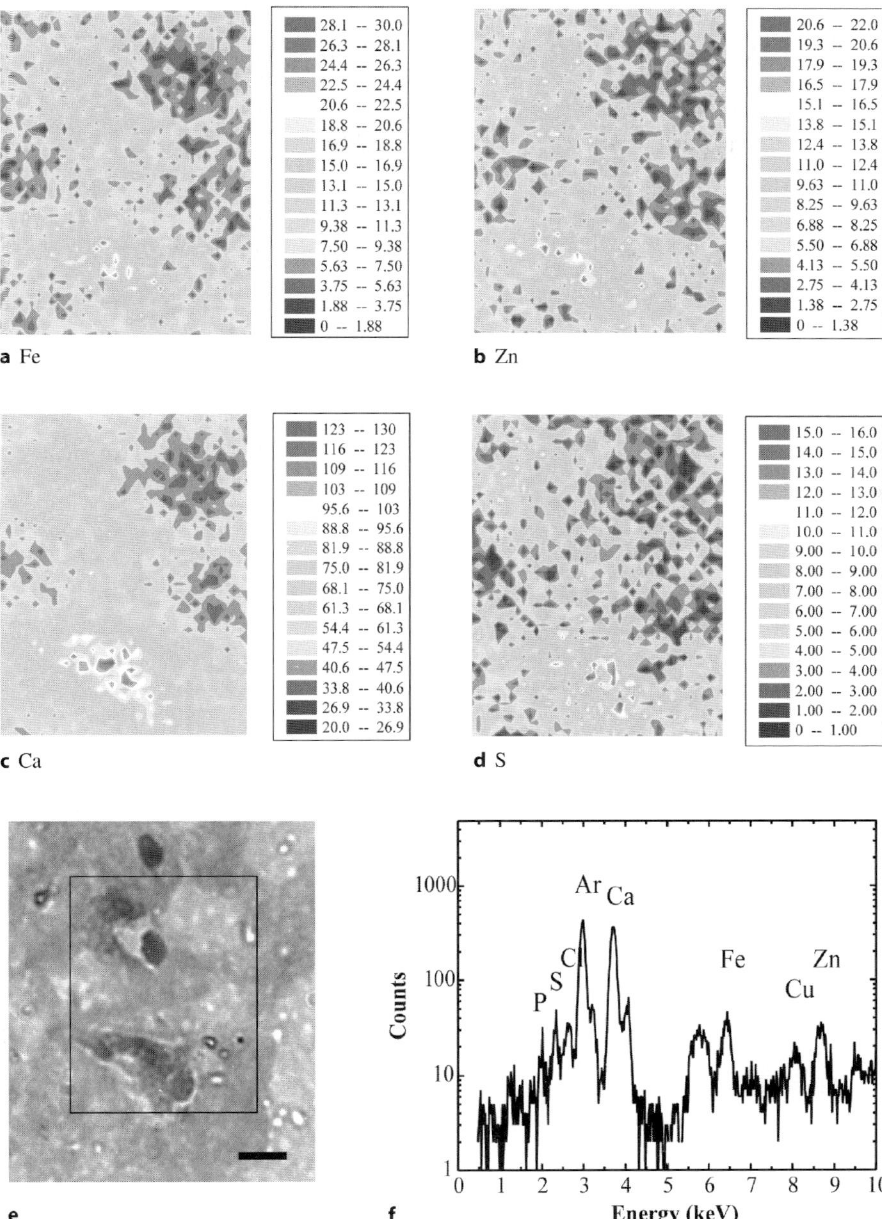

Fig. 7.32a–f. XRF images of Fe, Zn, Ca and S in the temporal lobes tissue of the AD patient (**a**)–(**d**). In this area, the tissue is not ubiquitin immunoreactive. The optical microscopic photograph of the sample stained with hematoxylin-eosin (HE) (**e**). The staining was performed after X-ray analysis. The scale bar is 20 μm. Typical X-ray fluorescence spectra in the tissue (**f**). The measurement time is 200 s

Table 7.3. Concentrations of elements contained in the tissues with Alzheimer's disease. The densities of the specimens are assumed to be $1.0\,\mathrm{g/cm^3}$. The unit is ppm. ub+: ubiquitin reactive. ub++: remarkable ubiquitin reactive

Patient	Cell	S	Ca	Fe	Cu	Zn	Ubiquitin
Patient 1(1)	A-1	4,585	9,879	328	80	319	ub+
	A-2	3,956	11,055	348	100	158	ub+
	A-3	4,511	10,406	352	88	143	ub+
	B-1-1	3,725	11,203	4,696	78	124	ub++
	B-1-2	7,074	17,007	1,241	586	1,227	ub++
	B-2	3,508	8,574	1,790	421	481	ub++
	C-1	3,163	11,910	4,025	211	812	ub++
	C-2	2,704	12,095	4,969	146	269	ub++
Patient 1(2)	A	6,528	13,896	454	78	144	
	B	6,766	15,862	1,168	76	168	
Patient 1(3)	A	4,336	12,939	811	73	244	
	B-1-1	4,948	14,306	653	260	267	
	B-1-2	5,458	11,974	611	1,524	379	
Patient 1(4)	A	2,202	36,558	856	55	643	
	B	2,651	5,466	184	58	38	
	C	2,534	5,453	2,102	74	49	
Patient 2	A	34,565	396,714	25,468	1,752	3,115	
	B	5,841	21,884	556	131	144	
	C	6,373	22,990	730	128	185	
	D	10,022	7,268	73,092	172	134	
Patient 3	A	2,232	2,988	30,677	44	82	
	B	2,136	10,305	120	57	72	
	C	1,836	2,431	13,101	53	85	
Patient 4	A	1,346	2,835	2,857	44	71	

order to analyze the chemical state of metallic elements in the AD tissues. Fe K-edge XANES analysis revealed that the Fe contained in the AD tissues shifted toward Fe^{3+}, indicating that the chemical state of iron contained in the tissue is highly oxidized.

The elevated level of iron in AD brain has been reported in other studies. For example, Ehmann et al. reported that iron level in the cortex of AD is $467\,\mu\mathrm{g/g}$, while that in control is $280\,\mu\mathrm{g/g}$ [50]. In this study by Fujisawa et al., the density of Fe in the tissues with AD was analyzed at the single cell level. The densities of Fe in the AD cortex were about $1000 \sim 4000\,\mathrm{ppm}$ in the neurons which contains abnormal Fe, and about $300 \sim 600\,\mathrm{pm}$ in the other neurons (Table 7.3). In the hippocampus tissues, abnormal concentration of Fe can be also observed. These results show the existence of neurons which have over 10 times higher Fe levels than normal neurons in AD brain. In this same study, XANES data of Fe in the AD tissue indicated a shift towards Fe^{3+}. This result indicates that Fe contained in the AD tissues may be at

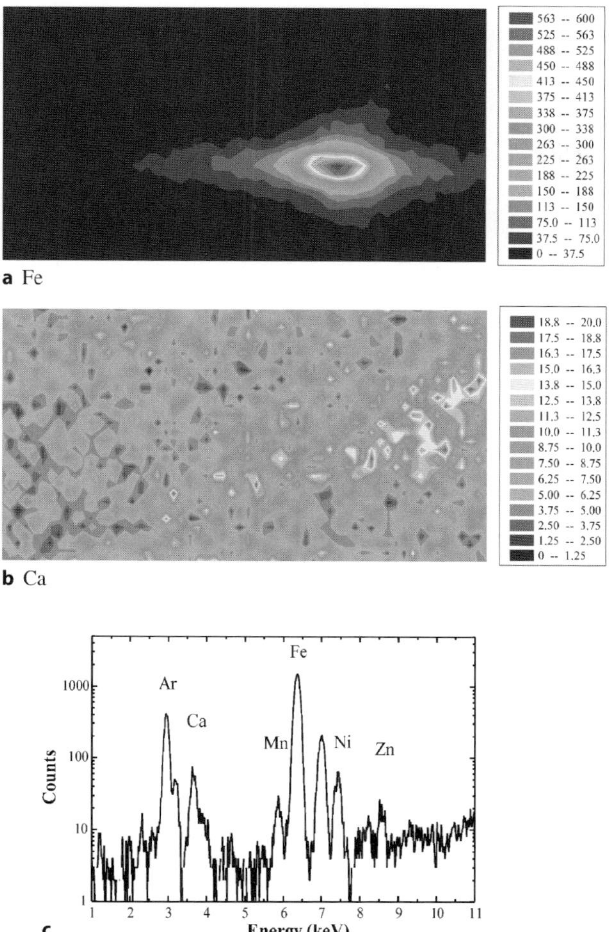

Fig. 7.33. (a,b) XRF images of Fe and Ca in the hippocampus tissue of the AD patient. (c) Typical X-ray fluorescence spectra in the tissue. The measurement time is 200 s

a higher oxidation state. With regard to the etiology of AD, iron accumulation is considered to contribute to the generation of oxidative species through Fenton reactions [51].

Fujisawa et al. also observed an increase of the densities of Cu and Zn in the ubiquitin deposited area, where there was a high correlation between Zn and Cu (the correlation coefficient was about 0.85). The densities of Zn and Cu in ubiquitin deposited tissues were also remarkably high in the section. This result suggests that zinc and copper may be correlated with the neuronal degeneration. In recent studies, it has been reported that Aβ binds copper and zinc to generate Cu/Zn SOD-like structure [52], which possesses catalytic

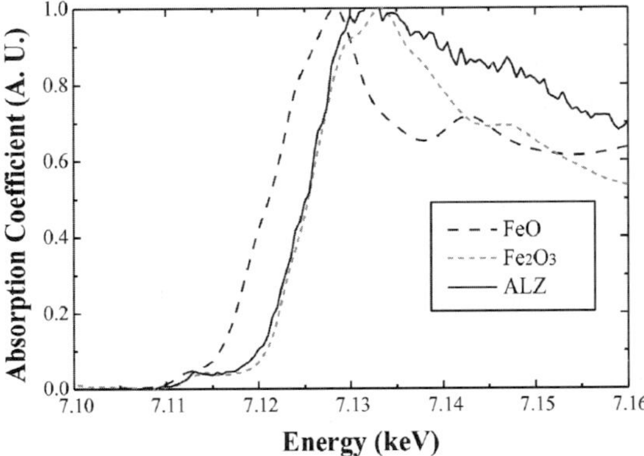

Fig. 7.34. Fe K-edge XANES spectra of the cell in the temporal lobes tissue of AD patient together with reference samples (FeO and Fe_2O_3 powder). The incident beam size was about 10 μm in diameter for the cell

Cu/Zn SOD (superoxid dismutase)-like activity [41]. In Fujisawa et al., the increase of the densities of Cu and Zn was observed in ubiquitin positive area, although they did not perform specific Aβ staining. The correlation between increase of Cu and Zn concentrations and the generation of deposit of Aβ, might be construed as the evidence that Aβ actually binds Cu/Zn (generating SOD-like structure) in AD brain.

7.5 Cell Degeneration in Friedreich Ataxia

7.5.1 Introduction

Friedreich's ataxia (FRDA) is a hereditary degenerative disease that involves the central and the peripheral nervous system and the heart. FRDA patient has mutations of a gene on chromosome 9, resulting in decrease of its expression [53]. The protein encoded by the FRDA gene, frataxin, is localized in a mitochondrion [54,55]. The functions of frataxin is still unknown, but is suggested to be relevant to the iron metabolism in the mitochondria from the experiments using the yeast frataxin homologue [56,57]. The deficiency of frataxin is suggested to cause iron accumulation in mitochondria, resulting in mitochondrial dysfunctions [58]. In FRDA patients the insufficient activity of iron-sulfur cluster-containing proteins and complexes, which is related to the respiratory chain reaction in mitochondria, can be observed [59]. It is supposed that iron accumulations in the mitochondria cause the production of free radicals, resulting in the deficiency of the iron-sulfur enzyme activity [59–61], because they are remarkably sensitive to the free radicals [62].

In FRDA patients, iron accumulations have been observed in the heart [63] and the dentate nucleus in cerebellum [64], which is the site of degeneration of the disease. It is supposed that there exists a relationship between iron accumulation in mitochondria and the deficiency of frataxin [60], because frataxin expressions are also tissue-specific in the human adult tissue [53] and in the mouse embryo [65]. This tissue-specificity is suggested to be due to the levels of respiratory activity (high in heart and CNS) and to the level of iron metabolism [60], but this is still unclear. Tissue specific vulnerability for oxidative stress induced by iron accumulation exists in other diseases such as dopaminergic neurons in substantia nigra with Parkinson's disease [66]. Iron metabolisms in different tissues with FRDA are important as the model of the tissue specific vulnerability for the free radical toxicity and the cell degeneration associated with iron overloads.

In order to analyze metallic elements, homogenization and histochemical methods such as staining are often used. However these methods are destructive and are difficult to apply for detection at trace levels. This is where SR-XRF with microbeams makes it possible to carry out elemental analysis at the single cell level.

Fujisawa et al. (2002) applied SR-XRF analysis to heart tissues with FRDA and a control case to investigate the distribution and the density of iron and other elements in the tissues, and to consider the free radical toxicity and the cell degeneration associated with iron overloads.

7.5.2 Experimental Procedures and Results

Fujisawa et al. used fixed heart tissues from a patient with Friedreich's ataxia and from a control. The tissues were embedded in paraffin and sections of $5 \sim 6\,\mu m$ in thickness were cut and mounted on PET film for X-ray analyses. Histochemical analyses were performed on parallel sections, by staining for collagen and iron deposition.

The SRXRF analyses were performed at the Photon Factory, KEK, using $14.3\,keV$ monochromated beam focused to $6 \times 5\,\mu m$. Incident and transmitted photon flux was monitored with an ion chamber. The fluorescent X-rays were collected by a solid state detector (SSD) and the measurements were performed in air.

The optical microscopic photograph of the parallel section with Fe staining and the specimen with no staining, and the XRF images of Fe and Ca are shown in Fig. 7.35. The cardiac muscle in parallel with the section can be observed. As can be seen in the XRF image, Fe was widely distributed in the certain myocardial cells in the measurement area. This result was in good agreement with the histochemical observation with the parallel section with Fe staining. The typical XRF spectra at the myocardial cell are shown in Fig. 7.36. As shown in the spectra, S, Ca, Cu and Zn were also present in the myocardial cell.

Corresponding results from heart tissues of the control case are shown in Fig. 7.37. The localization of Fe can be observed in a small area. This localization of Fe was not detected by iron staining in the parallel section.

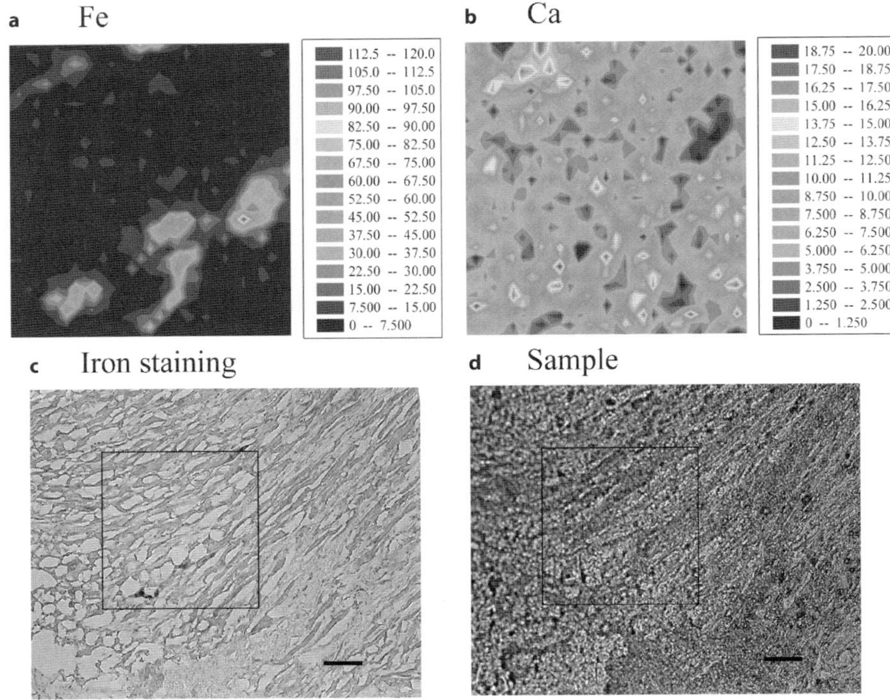

Fig. 7.35a–d. XRF imaging of Fe and Ca in the heart tissue from the patient with FRDA. Measurement area is $400 \times 400\,\mu m$, and the step is $10\,\mu m$. Measurement time was $3\,s$ per point. Scale bar is $100\,\mu m$

The typical XRF spectra at the myocardial cell are shown in Fig. 7.38. As shown in the spectra, S, Ca, Cu and Zn were also present in the cells.

Quantitative analyses of the concentrations of iron and other elements were performed at typical points in the tissues with FRDA and in the control case. The quantification results are shown in Table 7.4. The density of the sample was assumed to be $1.0\,g/cm^3$. In the control tissues, the density of Fe was 1.0×10^3 ppm at the highest point. And the density of S was 2.6×10^3 ppm at the same point. In the FRDA tissues, the density of Fe was 3.4×10^3 ppm at the highest point, which is about three times higher than that in the control tissues. At this point, the density of S was 4.0×10^3 ppm, which was also about 1.5 times higher than that in the control tissues.

7.5.3 Discussions and Summary

Fujisawa et al. (2002) used SR-XRF to study the distribution of iron and other elements and to quantify them in heart tissues from a patient with Friedreich ataxia (FRDA) and a control case, in order to investigate the free radical toxicity and cell degeneration associated with iron overload. Their

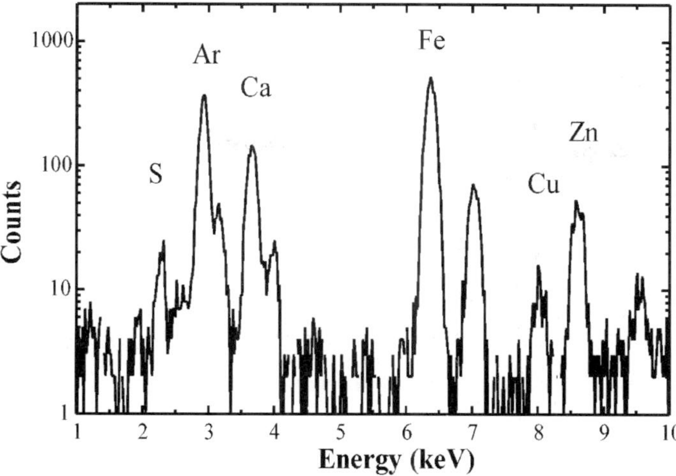

Fig. 7.36. Typical XRF spectra in the heart tissue from a patient with FRDA. Measurement time was 200 s

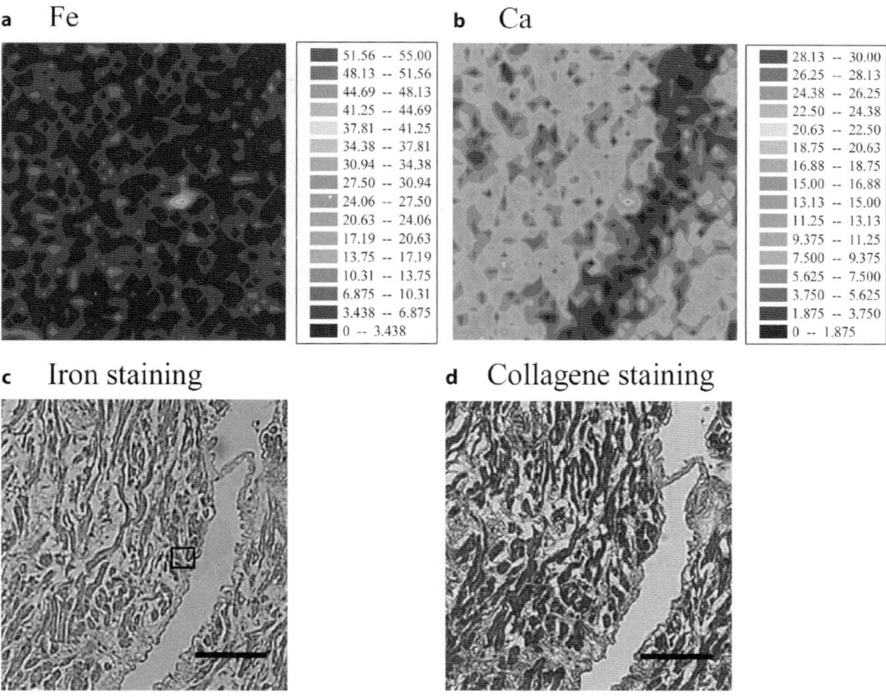

Fig. 7.37a–d. XRF imaging of Fe and Ca in the heart tissue from the patient with non-FRDA. Measurement area is $400 \times 400\,\mu m$, and the step is $10\,\mu m$. Measurement time was 3 s per point. Scale bar is $100\,\mu m$

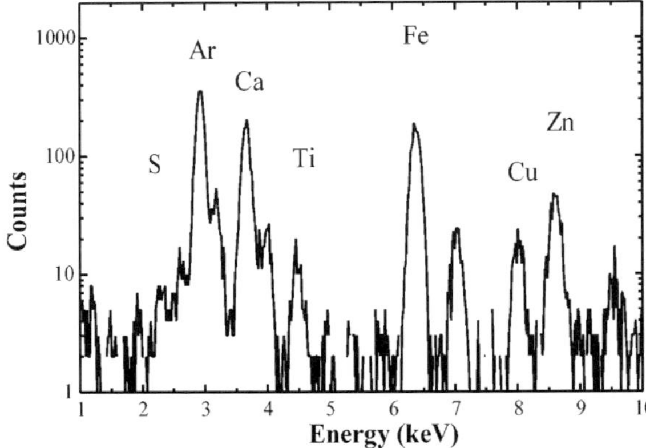

Fig. 7.38. Typical XRF spectra in the heart tissue from control. Measurement time was 200 s

Table 7.4. The density of the elements contained in the cells with FRDA and control. The values are the average of n points. The unit is ppm

	S	Ca	Fe	Cu	Zn	n
FRDA	4,038	4,844	3,440	59	213	5
Control	2,570	5,355	969	88	199	3

analysis results revealed excessive accumulation of iron in the FRDA tissues. The quantitative analysis showed that the density of Fe in the FRDA tissues was 3.4×10^3 ppm, which is about three times higher than that in the control tissues. The accumulation of iron in the heart of FRDA patients was also reported in previous studies [63]. Furthermore the density of S in the FRDA tissue also increased about two times higher than that in the control tissue. This result may suggest that the accumulation of iron-sulfur containing protein is caused as well as accumulation of iron in FRDA tissue. Rtig et al. reported the deficiency of the activity of iron-sulfur cluster-containing proteins in mitochondria with FRDA, indicating the free radical toxicity induced by excessive accumulation of iron. The relationship between the deficiency of the activity of the iron-sulfur protein and the increase of the iron and sulfur levels is obscure, but it may involve the iron metabolism in mitochondria with FRDA.

In this study, analysis of iron and other elements were applied to only heart tissues. However, the XRF technique using SR microbeams can be useful for other cellular elemental analysis, and help to investigate the relationship of iron metabolism between the different tissues.

References

1. L.P. Rowland, N.A. Shneider, *N. Engl. J. Med.*, **2001**, 344, 1688.
2. Y. Kokubo, S. Kuzuhara, Y. Narita, K. Kikugawa, R. Nakano, T. Inuzuka, S. Tsuji, M. Watanabe, T. Miyazaki, S. Murayama, Y. Ihara, *ARch. Neurol.*, **1999**, 56, 1506.
3. I. Fridovich, *Advances in Enzymology and related areas of molecular biology*, **1986**, 58, 61.
4. D.W. Cleveland, J. Liu, *Nature Med.*, **2000**, 6, 1320.
5. D.R. Rosen, T. Siddique, D. Patterson, D.A. Figlewicz, P. Sapp, A. Hentati, D. Donaldson, J. Goto, J.P. Oregan, H.X. Deng, Z. Rahmani, A. Krizus, D. Mckennayasek, A. Cayabyab, S.M. Gaston, R. Berger, R.E. Tanzi, J.J. Halperin, B. Herzfeldt, R. Vandenbergh, W.Y. Hung, T. Bird, G. Deng, D.W. Mulder, C. Smyth, N.G. Laing, E. Soriano, M.A. Pericakvance, J. Haines, G.A. Rouleau, J.S. Gusella, H.R. Horvitz, R.H. Brown, *Nature*, **1993**, 362, 59.
6. W. Robberecht, *J. Neurol.*, **2000**, 247, S1.
7. L.I. Bruijn, M.K. Houseweart, S. Kato, K.L. Anderson, S.D. Anderson, E. Ohama, A.G. Reaume, R.W. Scott, D.W. Cleveland, *Science*, **1998**, 281, 1851.
8. M.E. Gurney, H.F. Pu, A.Y. Chiu, M.C. Dalcanto, C.Y. Polchow, D.D. Alexander, J. Caliendo, A. Hentati, Y.W. Kwon, H.X. Deng, W.J. Chen, P. Zhai, R.L. Sufit, T. Siddique, *Science*, **1994**, 264, 1772.
9. J.P. Crow, J.B. Sampson, Y.X. Zhuang, J.A. Thompson, J.S. Beckman, *J. Neurochem.*, **1997**, 69, 1936.
10. A.G. Estevez, J.P. Crow, J.B. Sampson, C. Reiter, Y.X. Zhuang, G.J. Richardson, M.M. Tarpey, L. Barbeito, J.S. Beckman, *Science*, **1999**, 286, 2498.
11. G.A. Rouleau, A.W. Clark, K. Rooke, A. Pramatarova, A. Krizus, O. Suchowersky, J.P. Julien, D. Figlewicz, *Ann. Neurol.*, **1996**, 39, 128.
12. L.M. Murphy, R.W. Strange, S.S. Hasnain, *Structure*, 1997, 5, 371.
13. P.J. Hart, H.B. Liu, M. Pellegrini, A.M. Nersissian, E.B. Gralla, J.S. Valentine, D. Eisenberg, *Protein Sci.*, **1998**, 7, 545.
14. H. Takahashi, *No to Shinkei*, **1972**, 47, 292.
15. S. Kato, H. Hayashi, K. Nakashima, E. Namba, M. Kato, A. Hirano, I. Nakano, K. Asayama, E. Ohama, *Am. J. Pathol.*, **1997**, 151, 611.
16. T. Koide, S. Igarashi, K. Kikugawa, R. Nakano, T. Inuzuka, M. Yamada, H. Takahashi, S. Tsuji, *Neurosci. Lett.*, **1998**, 257, 29.
17. *"Molecular Mechanism and Therapeutics of Amyotrophic Lateral Sclerosis"*, ed. K Abe, **2001**, 41.
18. P.J. Shaw, V. Forrest, P.G. Ince, J.P. Richardson, H.J. Wastell, *Neurodegeneration,* **1995**, 4, 209.
19. Y. Ogawa, H. Kosaka, T. Nakanishi, A. Shimizu, N. Ohoi, H. Shouji, T. Yanagihara, S. Sakoda, *Biochem. Biophys. Res. Commun.*, **1997**, 241, 251.
20. A.G. Butt, J.M. Bowler, C.W. McLaughlin, *J. Membr. Biol.*, **1998**, 162, 17.
21. G. Weiss, G. Werner-Felmayer, E.R. Werner, K. Grunewald, H. Wachter, M.W. Hentze, *J. Exp. Med.*, **1994**, 180, 969.
22. P.J. Shaw, R.M. Chinnery, P.G. Ince, *Brain Res.*, **1994**, 655, 186.
23. S.G. Carriedo, H.Z. Yin, J.H. Weiss, *J. Neurosci.*, **1996**, 16, 4069.
24. S. Kwak, R. Nakamura, *J. Neurol. Sci.*, **1995**, 129, 99.
25. Y.M. Lu, H.Z. Yin, J. Chiang, J.H. Weiss, *J. Neurosci.*, **1996**, 16, 5457.

26. S.G. Carriedo, H.Z. Yin, S.L. Sensi, J.H. Weiss, *J. Neurosci.*, **1998**, 18, 7727.
27. S.G. Carriedo, S.L. Sensi, H.Z. Yin, J.H. Weiss, *J. Neurosci.*, **2000**, 20, 240.
28. M. Kondo, R. Sumino, H. Okado, *J. Neurosci.*, **1997**, 17, 1570.
29. M.E. Alexianu, B.K. Ho, A.H. Mohamed, V. La Bella, R.G. Smith, S.H. Appel, *Ann. Neurol.*, **1994**, 36, 846.
30. A. Takahashi, P. Camacho, J.D. Lechleiter, B. Herman, *Phys. Rev.*, **1999**, 79, 1089.
31. J.W. McDonald, T. Bhattacharyya, S.L. Sensi, D. Lobner, H.S. Ying, L.M. Canzoniero, D.W. Choi, *J. Neurosci.*, **1998**, 18, 6290.
32. S.L. Sensi, H.Z. Yin, S.G. Carriedo, S.S. Rao, J.H. Weiss, *Neurobiol.*, **1999**, 96, 2414.
33. D.J. Selkoe, *Trends. Neurosci.*, **1993**, 16, 403.
34. A.I. Bush, W.H. Pettingell, M.D. Paradis, R.E. Tanzi, *J. Biol. Chem.*, **1994**, 269, 12152.
35. C.S. Atwood, R.C. Scarpa, X. Huang, R.D. Moir, W.D. Jones, D.P. Fairlie, R.E. Tanzi, A.I. Bush, *J. Neurochem.*, **2000**, 75, 1219.
36. R.A. Cherny, C.S. Atwood, M.E. Xilinas, D.N. Gray, W.D. Jones, C.A. McLean, K.J. Barnham, I. Volitakis, F.W. Fraser, Y.S. Kim, X. Huang, L.E. Goldstein, R.D. Moir, J.T. Lim, K. Beyreuther, H. Zheng, R.E. Tanzi, C.L. Masters, A.I. Bush, *Neuron*, **2001,** 30, 665.
37. A.I. Bush, W.H. Pettingell, G. Multhaup, M.D. Paradis, J.P. Vonsattel, J.F. Gusella, K. Beyreuther, C.L. Masters, R.E. Tanzi, *Science*, **1994**, 265, 1464.
38. R.A. Cherny, J.T. Legg, C.A. McLean, D.P. Fairlie, X.D. Huang, C.S. Atwood, K. Beyreuther, R.E. Tanzi, C.L. Masters, A.I. Bush, *J. Biol. Chem.*, **1999**, 274, 23223.
39. X.D. Huang, M.P. Cuajungco, C.S. Atwood, M.A. Hartshorn, J.D. Tyndall, G.R. Hanson, K.C. Stokes, M. Leopold, G. Multhaup, L.E. Goldstein, R.C. Scarpa, A.J. Saunders, J. Lim, R.D. Moir, C. Glabe, E.F. Bowden, C.L. Masters, D.P. Fairlie, R.E. Tanzi, A.I. Bush, *J. Biol. Chem.*, **1999**, 4, 37111.
40. X.D. Huang, C.S. Atwood, M.A. Hartshorn, G. Multhaup, L.E. Goldstein, R.C. Scarpa, M.P. Cuajungco, D.N. Gray, J. Lim, R.D. Moir, R.E. Tanzi, A.I. Bush, *Biochemistry,* **1999**, 38, 7609.
41. A.I. Bush, *Curr. Opin. Chem. Biol.*, **2000**, 4, 184.
42. C.M. Thompson, W.R. Markesbery, W.D. Ehmann, Y.X. Mao, D.E. Vance, *Neurotoxicology,* **1988**, 9, 1.
43. J.R. Connor, S.L. Menzies, S.M. St. Martin, E.J. Mufson, *J. Neurosci. Res.*, **1992**, 31, 75.
44. M.A. Smith, P.L. Harris, L.M. Sayre, G. Perry, *Proc. Natl. Acad. Sci. USA,* **1997**, 94, 9866.
45. M.A. Smith, K. Wehr, P.L. Harris, S.L. Siedlak, J.R. Connor, G. Perry, *Brain Res.*, **1998**, 788, 232.
46. Q. Guo, B.L. Sopher, K. Furukawa, D.G. Pham, N. Robinson, G.M. Martin, M.P. Mattson, *J. Neurosci.*, **1997**, 17, 4212.
47. M.P. Mattson, H.Y. Zhu, J. Yu, M.S. Kindy, *J. Neurosci.*, **2000**, 20, 1358.
48. A. Alves-Rodrigues, L. Gregori, M.E. Figueiredo-Pereira, *Trends. Neurosci.*, **1998**, 21, 516.
49. T. Kudo, K. Iqbal, R. Ravid, D.F. Swaab, I. Grundkeiqbl, *Brain Res.*, **1994**, 639, 1.

50. W.D. Ehmann, W.R. Markesbery, M. Alsuddin, T.I. Hossain, E.H. Brubaker, *Neurotoxicology,* **1986**, 7, 197.

51. W.R. Markesbery, *Free Radical Bio. Med.,* **1997**, 23, 134.

52. C.C. Curtain, F. Ali, I. Volitakis, R.A. Cherny, R.S. Norton, K. Beyreuther, C.J. Barrow, C.L. Masters, A.I. Bush, K.J. Barnham, *J. Biol. Chem.,* **2001**, 276, 20466.

53. V. Campuzano, L. Montermini, M.D. Molto, L. Pianese, M. Cossee, F. Cavalcanti, E. Monros, F. Rodius, F. Duclos, A. Monticelli, F. Zara, J. Canizares, H. Koutnikova, S.I. Bidichandani, C. Gellera, A. Brice, P. Trouillas, G. DeMichele, A. Filla, R. DeFrutos, F. Palau, P.I. Patel, S. DiDonato, J.L. Mandel, S. Cocozza, M. Koenig, M. Pandolfo, *Science,* **1996**, 271, 1423.

54. V. Campuzano, L. Montermini, Y. Lutz, L. Cova, C. Hindelang, S. Jiralerspong, Y. Trottier, S.J. Kish, B. Faucheux, P. Trouillas, F. J. Authier, A. Durr, J.L. Mandel, A. Vescovi, M. Pandolfo, M. Koenig, *Hum. Mol. Genet.,* **1997**, 6, 1771.

55. M. Cossée, V. Campuzano, H. Koutnikova, K. Fischbeck, J.L. Mandel, M. Koenig, S.I. Bidichandani, P.I. Patel, M.D. Molte, J. Canizares, R. DeFrutos, L. Pianese, F. Cavalcanti, A. Monticelli, S. Cocozza, L. Montermini, M. Pandolfo, *Nat. Genet.,* **1997**, 15, 337.

56. M. Babcock, D. De Silva, R. Oaks, S. Davis-Kaplan, S. Jiralerspong, L. Montermini, M. Pandolfo, J. Kaplan, *Science,* **1997**, 276, 1709.

57. D.C. Radisky, M.C. Babcock, J. Kaplan, *J. Biol. Chem.,* **1999**, 274, 4497.

58. M.B. Delatycki, J. Camakaris, H. Brooks, T. Evans-Whipp, D.R. Thorburn, R. Williamson, S.M. Forrest, *Ann. Neurol.,* **1999**, 45, 673.

59. A. Rötig, P. De Lonlay, D. Chretien, F. Foury, M. Koenig, D. Sidi, A. Munnich, P. Rustin, *Nat. Genet.,* **1997**, 17, 215.

60. M. Pandolfo, *Arch. Neurol.,* **1999**, 56, 1201.

61. E. Becker, D.R. Richardson, *Int. J. Biochem. Cell B,* **2001**, 33, 1.

62. I. Fridovich, *Annu. Rev. Biochem.,* **1995**, 64, 97.

63. G. Sachez-Casis, M. Cote, A. Barbeau, *Can. J. Neurol. Sci.,* **1976**, 3, 349.

64. D. Waldvogel, P. van Gelderen, M. Hallett, *Ann. Neurol.,* **1999**, 46, 123.

65. S. Jiralerspong, Y. Liu, L. Montermini, S. Stifani, M. Pandolfo, *Neurobiol. Dis.,* **1997**, 4, 103.

66. E.C. Hirsch, *Ann. Neurol.,* **1992**, 32, S88.

67. P.L. McGeer, S. Itagaki, H. Akiyama, E.G. McGeer, *Ann. Neurol.,* **1988**, 24, 574.

SR Analysis of Tissues

8.1 Quantification Analysis of Zinc in Prostate Cancer Tissues

8.1.1 Introduction

A human prostate gland has a unique feature of possessing high level accumulation of zinc compared to other organs [1]. In prostate cancer tissue, this unique zinc metabolism is altered resulting in the decrease of zinc level in the epithelial cells [1,2]. Zinc accumulation in glandular cells is relevant to the citrate metabolism in the citric acid cycle (Krebs cycle). The citric acid cycle is a series of biochemical reactions, which involve the energy metabolism in the cells [3]. Via the cycle citrate is oxidized and isomerized into isocitrate by an enzyme, aconitase. Zinc inhibits the activity of aconitase, resulting in inhibition of the citrate oxidation [4]. Citrate is also abundant in a prostate gland, which has the unique function of secretion [5]. A high level of zinc may involve the regulation of citrate oxidation in a prostate gland. In a prostate cancer tissue, citrate is remarkably decreased as well as the level of zinc. The question arises as how the reduction of zinc level involves the progression of malignancy in prostate cancer tissue. As one possibility, Costello et al. have suggested that excessive citrate oxidation result in production of coupled ATP, which is essential for the progression of malignancy [6,7]. On the other hand, a hormone is closely related with the zinc level and malignancy in a prostate tissue. Inhibition of phosphatidylinositol 3-kinase induces cellular apoptosis. Androgen inhibits this pathway, i.e., androgen inhibits apoptosis [8]. Zn^{2+} induces stimulation of phosphoinositide 3-kinase [9], thereby zinc inhibits apoptosis. Apoptosis is a mechanism of programmed death of cell, and deficient apoptosis is related to the progression of malignancy.

Thus the role of zinc in normal and cancerous prostate tissues presents an interesting challenge. Zinc may have regulatory function in the cellular biochemical reactions, such as in the citric acid cycle, and it would be of interest to investigate the relationship of the zinc behavior and other biochemical species or elements. SR-XRF spectroscopy provides the opportunity by enabling multi-elemental analysis at the cell level.

Fujisawa et al. (2002) conducted a series of studies, where they applied SR-XRF analysis to human prostate tissues (normal and cancerous tissues) and cultured prostate cancerous cells, in order to consider the role of zinc as a regulatory factor.

8.1.2 Zinc Distribution in Human Prostate Cancer Tissues and Normal Tissues

In their first study, Fujisawa et al. (2002) applied XRF analysis to both cancerous and control tissues of the human prostate. The aim of that study was to investigate the differences in the densities and the distributions of zinc and other elements and to show the possibilities for applications of this technique to clinical diagnosis. They proposed a new method, called the "statistical correlation diagnosis method" to differentiate between the normal and diseased tissues.

The samples were specimens obtained from the prostate gland of a 56-year-old male patient, who was diagnosed with adenocarcinoma of the prostate. The sample tissues were a mixture of cancerous and normal tissues, and were prepared by fixing in 4% formalin and embedding in paraffin. Unstained sections with a 5 μm thickness were cut from the paraffin block and mounted on a nylon film for the XRF analyses. Parallel sections of the specimen were stained by hematoxylin and eosin (HE) staining for optical microscopic study.

The SR-XRF analyses were performed at the Photon Factory (KEK) using monochromated 14.3 keV X-rays with beam size of $6 \times 5 \, \mu m^2$. The fluorescent X-rays were collected by a solid state detector (SSD) and the measurements were performed in air.

XRF analyses were performed in the cancerous prostate tissues. The photograph of the parallel sections of the tissues (approximate measurement area) stained by HE is shown in Fig. 8.1a. XRF images of Zn and Ca (in the wide area of the tissue) are shown in Fig. 8.1b,c, respectively. XRF images have matrices of 40×40 pixels at a 5 μm resolution, and the X-ray fluorescent yields in each pixel are represented by a color scale. The measurement time was 4 seconds for each pixel. The measurement area was selected to include the adenocarcinoma (the cancerous epithelial tissues), where zinc and calcium were detected. A high concentration of Zn and Ca were observed at some points in the stroma, shown in Fig. 8.1d,e, respectively. XRF images have matrices of 40×40 pixels at a 2 μm resolution. The measurement time was 5 seconds for each pixel. The typical XRF spectra at the adenocarcinoma and at the stroma are shown in Fig. 8.2. As shown in these spectra, Fe, Cu and S were also present in the cancer cells.

The same analyses were performed in the normal prostate tissues of the same patient. The photograph of the parallel sections of the tissues (approx-

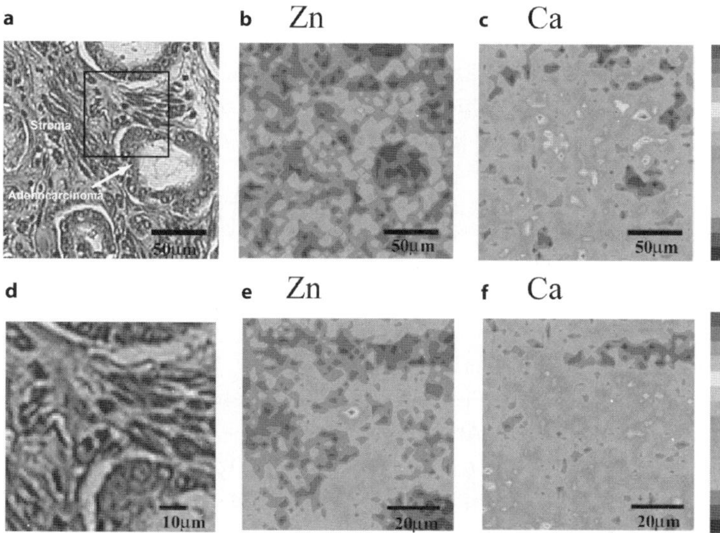

Fig. 8.1a–f. XRF images in the prostate cancer tissue. (**a** and **d**) The optical microscopic photograph of the measurement area. (**b**) XRF imaging of Zn (color scale 0–3.3×10^2 ppm). (**c**) XRF image of Ca (color scale 0–3.2×10^3 ppm). (**b**)–(**c**) The measurement area is $200 \times 200\,\mu$m, and the step is $5\,\mu$m. (**e**) Detailed XRF image of Zn (color scale 0–5.3×10^2 ppm). (**f**) Detailed XRF image of Ca (color scale 0–3.3×10^3 ppm). The measurement area is shown in (**a**) by a square ($80 \times 80\,\mu$m). The step was $2\,\mu$m

imate measurement area) stained by HE is shown in Fig. 8.3a, and the XRF images of Zn and Ca (in the wide area of the tissue) are shown in Fig. 8.3b,c, respectively. XRF images have matrices of 40×40 pixels at a $5\,\mu$m resolution, and the X-ray fluorescent yields in each pixel are represented by a color scale. The measurement time was 3 seconds for each pixel. The measurement area was selected to include the normal epithelial tissues. The distributions of Zn and Ca can be observed in normal epithelial cells and in the stroma of prostate. The detailed XRF image of Zn and Ca in one of the cells in this area are shown in Fig. 8.3d and e, respectively. XRF images have matrices of 40×40 pixels at a $1\,\mu$m resolution. The measurement time was 5 seconds for each pixel. A typical XRF spectrum in the normal tissues is shown in Fig. 8.4. Cu and Fe are also detected in the normal cells.

Results of quantification analyses of the elements in the cancer tissue and normal tissue are shown in Table 8.1. The density of the sample was assumed to be $1.0\,\mathrm{g/cm^3}$. In the cancer tissue, zinc was detected at the adenocarcinoma and the stroma of the prostate. The density of Zn was $89 \sim 221$ ppm in the adenocarcinoma and $44 \sim 713$ ppm in the cells in the stroma. In particular, high level of Zn was detected at some points in the stroma. In the normal tissues, Zn was detected at $158 \sim 474$ ppm in the epithelial cells, and $101 \sim$

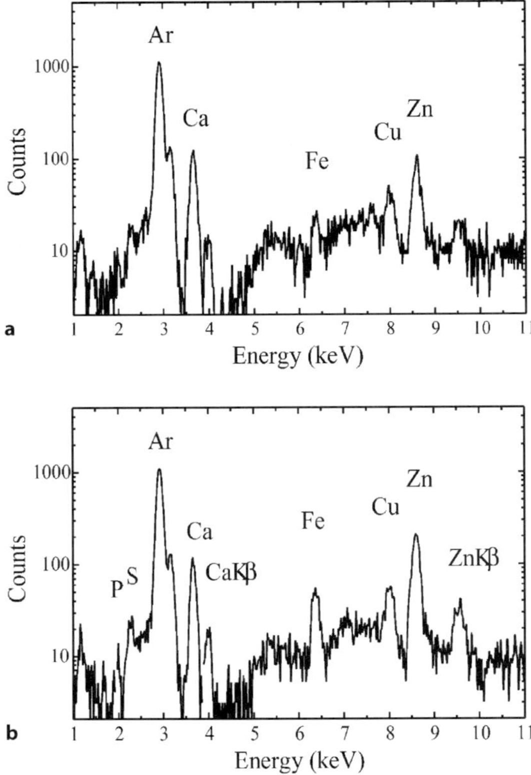

Fig. 8.2a,b. Typical XRF spectra in the prostate cancer tissues. (**a**) In the adenocarcinoma. (**b**) In the stroma

Table 8.1. The density of Zn and Ca contained in the prostate tissues (in ppm). The values in the brackets are the average of data at n points

Tissues	Cell type	Zn	Ca	n
Cancer	Adenocarcinoma	89 − 221 (149.2)	1.4 − 2.3E3 (1.83E3)	14
	Stroma	44 − 713 (184.1)	0.8 − 2.6E3 (1.61E3)	34
Normal	Epithelial cells	158 − 474 (290.5)	2.5 − 17.6E3 (6.54E3)	14
	Stroma	101 − 180 (140.7)	2.7 − 7.2E3 (3.87E3)	8

180 ppm in the stroma. Calcium was also detected in the cancer tissues as well as in the normal tissues. In the cancer tissues, the density of Ca in the adenocarcinoma and the stroma was $1.4 \sim 2.3 \times 10^3$ and $0.8 \sim 2.6 \times 10^3$ ppm, respectively. In the normal tissues, the density of the calcium in the epithelial cells and in the stroma was $2.5 \sim 17.5 \times 10^3$ and $2.7 \sim 7.2 \times 10^3$ ppm, respectively.

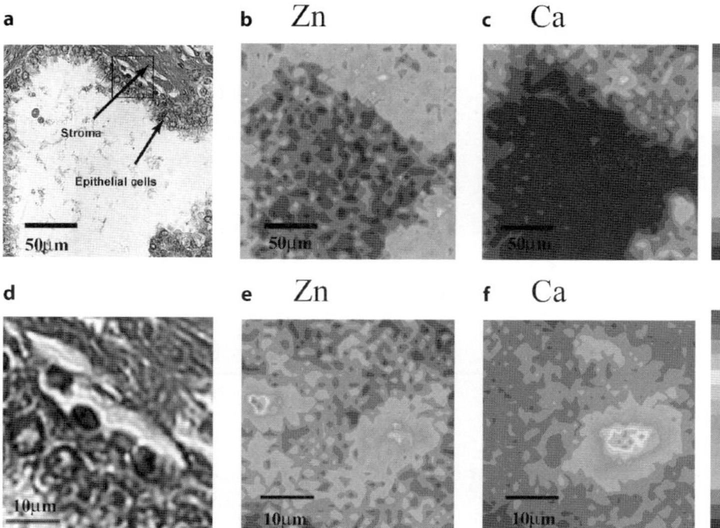

Fig. 8.3a–f. XRF images in the normal prostate tissue. (a and d) The optical microscopic photograph of the measurement area. (b) XRF image of Zn (color scale 0–5.3×10^2 ppm), (c) XRF image of Ca (color scale 0–1.8×10^4 ppm). (b)–(c) The measurement area is $200 \times 200\ \mu$m, and the step is $5\ \mu$m. (e) Detailed XRF image of Zn (color scale 0.3–4.4×10^2 ppm), (f) detailed XRF image of Ca (color scale 0–1.4×10^4 ppm). The measurement area is shown in (a) by a square ($40 \times 40\ \mu$m). The step was $1\ \mu$m

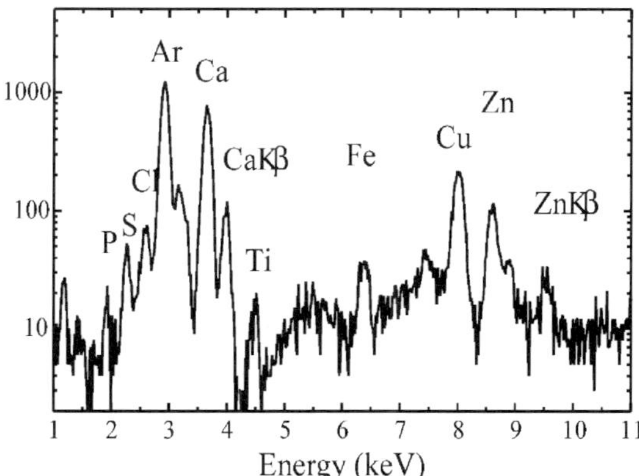

Fig. 8.4. Typical XRF spectrum at the epithelial cells in the normal tissues

In addition to the imaging and quantification of the elements that gave detailed information of the cells, their constituent elements, and the variations in the density distribution of the elements, Fujisawa et al. (2002) also developed a new technique to perform diagnosis based on a statistical study of the cells in the cancerous and the normal tissues. This method, which they

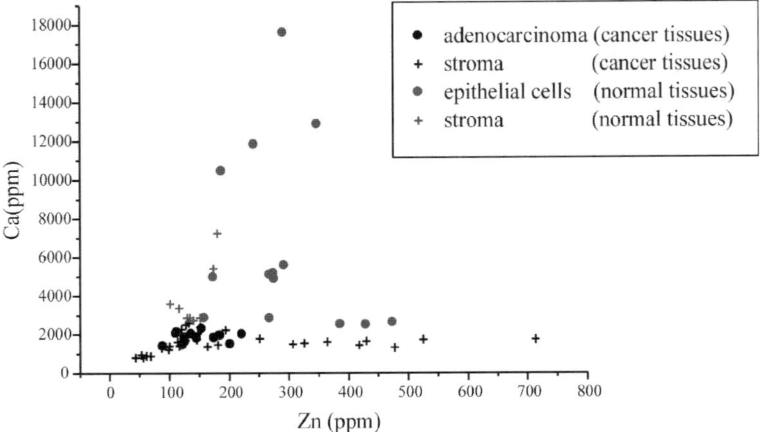

Fig. 8.5. The correlation of the density of Zn and Ca contained in the prostate tissue

called the "statistical correlation diagnosis method", is performed according to the following procedures: (i) first the densities of the elements from each spectrum obtained at certain points in the cells are quantified, then (ii) the x and y axes are taken as the density of Zn and Ca, respectively, and finally (iii) the quantified values of the correlation between Zn and Ca are plotted.

The correlation of the densities of Zn and Ca are shown in Fig. 8.5. In this figure, the cancer tissues are divided into two groups (adenocarcinoma and stroma), and the normal cells are divided into two groups (epithelial cells and stroma). Significant differences in the density correlation can be observed between the cancerous tissues and the normal tissues. In the normal tissues, the densities of Zn and Ca have a high fluctuation and the regression coefficient is 4.91. However, on the contrary, in the case of the cancerous tissues, the density of Ca does not have such a fluctuation and the regression coefficient is 0.29.

In the second experiment, Fujisawa et al. (2002) analyzed specimens from the prostate gland of patients who were diagnosed with adenocarcinoma of the prostate, and of patients who were diagnosed with benign prostatic hypertrophy (BPH), looking for the difference of elemental distributions in the tissues such as neoplasia and the stroma. The experiment was carried out using high flux beam, restricted bay a pinhole in a tantalum disk (φ 2.4 µm), at BL40XU in SPring-8, allowing fast data acquisition. The transmitted X-rays were monitored by a PIN photodiode, and the fluorescent X-rays were collected by a Si drift detector (max 30,000 cps). The sample stage has $x - y$ motion on vertical plane against the beam, driven by stepping motors. A CCD camera is equipped in front of the sample holder, and a sample can be monitored during a measurement.

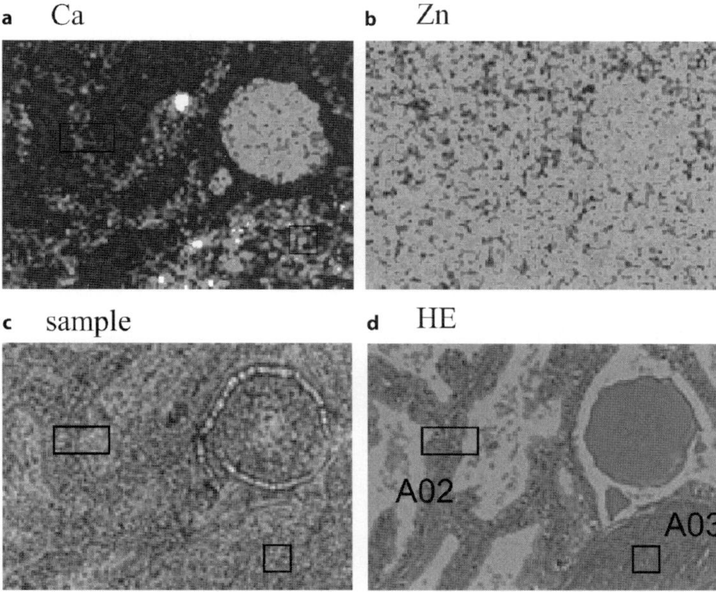

Fig. 8.6a–d. Elemental images of the tissue with benign prostatic hypertrophy (BPH)

As in the first experiment, the specimens were fixed in 4% formalin and embedded in paraffin. Unstained sections with a 5 μm thickness were cut from the paraffin block and mounted on a nylon film for the XRF analyses. Parallel sections of the specimen were stained by hematoxylin and eosin (HE) staining for optical microscopic study.

XRF analyses were applied to a BPH and two prostate cancerous tissues (low and high grade of malignancy), in order to obtain the features of distributions of zinc in these tissues. First, XRF analysis was performed to the BPH tissues. XRF imaging of Ca and Zn are shown in Fig. 8.6a,b, respectively. The photograph of the sample section and the parallel sections of the tissues (approximate measurement area) stained by HE is shown in Fig. 8.6c,d. XRF images have matrices of 100×70 pixels at a 5 μm resolution, and the measurement time was 1 second for each pixel. Distribution of Zn was diffuse in the BPH tissue, and was similar between in the epithelial neoplasia area and in the stroma. The excessive accumulation of zinc was not observed in the BPH tissue. On the other hand, calcium accumulation was observed at one point in the neoplasia and at some points in the stroma. The distribution of Ca well reflects the morphology of the BPH tissue.

Subsequently XRF analysis was performed to the prostate cancerous tissues with low grade of malignancy. XRF images of Zn and Ca are shown in Fig. 8.7a,b, respectively. The photograph of the parallel sections stained by HE is shown in Fig. 8.7c. XRF images have matrices of 50×70 pixels at a 5 μm resolution, and the measurement time was 2 seconds for each pixel.

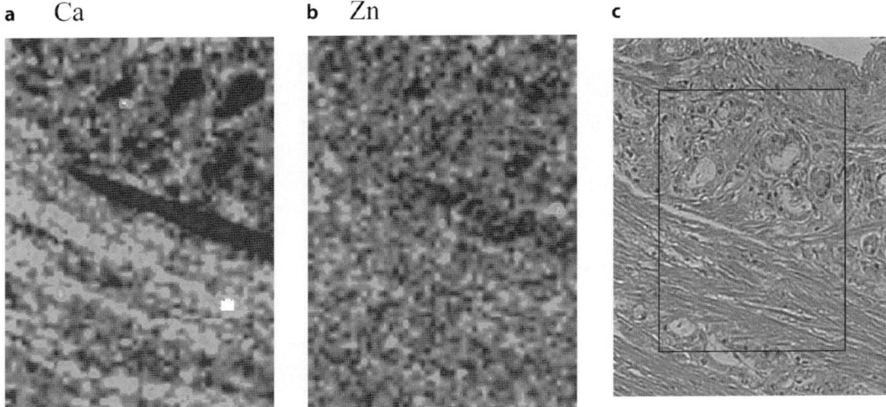

Fig. 8.7a–c. Elemental image of the tissue with low grade of malignancy

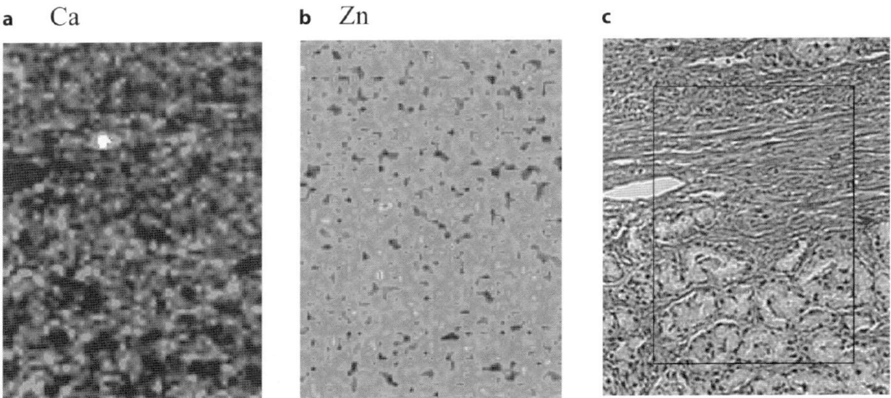

Fig. 8.8a–c. Elemental images of the tissue with high grade of malignancy

The distribution of zinc in the cancerous tissue was similar to that of BPH. Calcium accumulation was observed in the stroma, but not in the adenocarcinoma tissues.

XRF analysis is also applied to the prostate cancerous tissues with high malignancy. XRF images of Ca and Zn (in the wide area of the tissue) are shown in Fig. 8.8a,b, respectively. The photograph of the parallel sections stained by HE is shown in Fig. 8.8c. XRF images have matrices of 50 × 70 pixels at a 5 μm resolution, and the measurement time was 2 seconds for each pixel. The distribution of zinc was observed in the boundary area of the adenocarcinoma and the stroma. The distribution of calcium in the high malignant cancerous tissue was similar to that in the middle malignant cancerous tissue.

The statistical correlation method mentioned previously was also applied to these tissues. The correlation of fluorescent yields of Zn and Ca are shown

Table 8.2. Summary of the characteristics of the prostate cancerous cell lines

	Androgene	Aggressive	Carcinoma	Metastatic site
PC-3	Independent	Yes	Prostate	Bone
DU145	Independent	Yes	Prostate	Brain
LNCaP	Dependent	No	Prostate	Lymph node

in Fig. 8.9. The fluorescent yield of each element was normalized by the yield of argon. In this figure, the cancer tissues were divided into two groups (adenocarcinoma and stroma), and the BPH tissues were divided into two groups (epithelial neoplasia cells and stroma). In the BPH tissues, the distribution of the correlation in the neoplasia and the stroma were similar. In the cancerous tissues with low malignancy, the distribution in the adenocarcinoma and the stroma were also similar, but the fluctuations of Ca density were reduced compared to the BPH case. In the cancerous tissues with high malignancy, the difference of the distribution between adenocarcinoma and stroma was observed. The average of the zinc density in the stroma was about 17% lower than that of adenocarcinoma. This result indicates that zinc reduction in the tissue with prostate tissue occurs in the stroma rather than in the adenocarcinoma.

In the third experiment, Fujisawa et al. (2002) focused on the relationship between the hormone resistance and the zinc. Androgen promotes the activity of the cancerous cells, hence androgen deprivation is one of the effective therapies for prostate cancer. However the adenocarcinoma of the prostate after hormonal manipulation often acquires resistance to the deprivation of androgen. The mechanism of the acquisition of the hormone resistance is unknown. They investigated the distribution and the density of zinc in the human prostate cancer cell lines PC-3 and DU145, which have hormone resistance, and LNCaP, which is hormone-dependence. The specimen used in that study was the human malignant cell lines PC-3, DU145, LNCaP and C4-2. The features of these cell lines are shown in Table 8.2. LNCaP is sensitive to hormone [10], while PC-3 [11] and DU145 [12] are independent of androgen. The experiment was carried out at the same facility as in the first experiment, namely BL4A at the Photon Factory.

XRF analyses were performed in the heart tissues with DU145 cells, and the images of zinc, calcium and sulfur are shown in Fig. 8.10a–c. The optical microscopic photograph of the parallel section with HE staining is shown in Fig. 8.10d. XRF images (36×33 pixels of $5\,\mu m$ size) were measured at 3 seconds for each pixel. Zinc concentrations can be observed in some of cells. The typical XRF spectra at the DU145 cell containing Zn are shown in Fig. 8.10e. Ca, Fe and Cu are also detected in the cell.

XRF images (40×40 pixels of $5\,\mu m$ size) of Zn, Ca and S in LNCaP cells are shown in Fig. 8.11a–c, measured at 3 seconds for each pixel. In these cells the Zn is detected low level as well as Ca. The typical XRF spectra at the LNCaP cell containing Zn are shown in Fig. 8.11e.

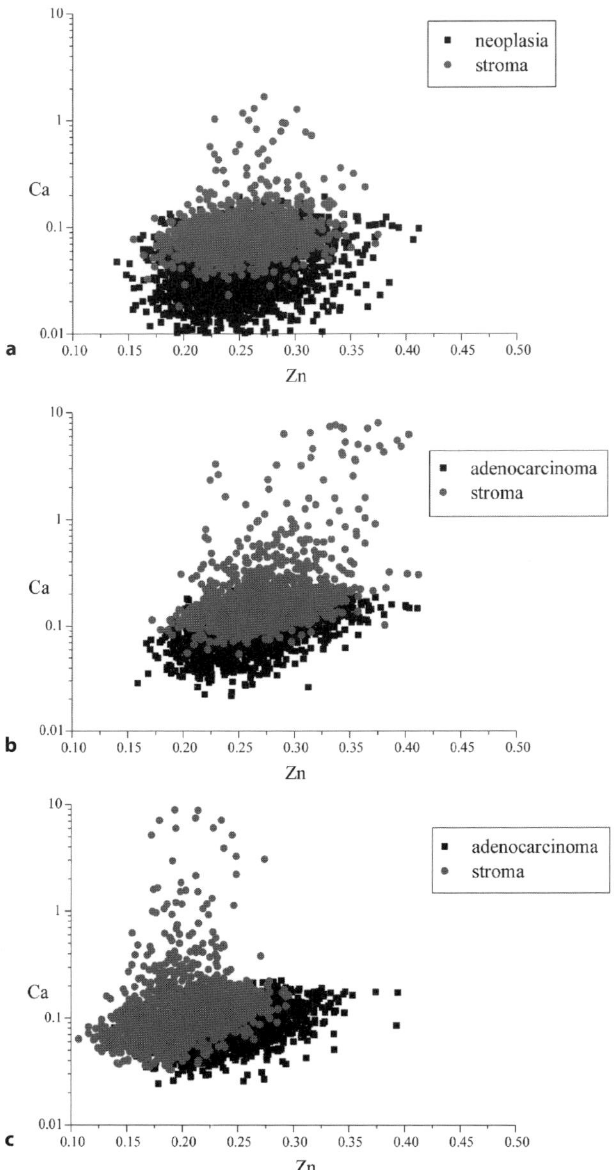

Fig. 8.9a–c. The correlation of the density of Zn and Ca contained in the prostate tissue with (**a**) BPH, (**b**) low grade of malignancy and (**c**) high grade of malignancy. The X-ray fluorescent yields of Zn and Ca are standardized by the yields of Ar

XRF images of Zn, Ca and S in PC-3 are shown in Fig. 8.12a–c. XRF images have matrices of 40×40 pixels of $5\,\mu$m size, and the measurement time was 3 seconds for each pixel. The typical XRF spectra at the PC-3 cell containing zinc are shown in Fig. 8.12e.

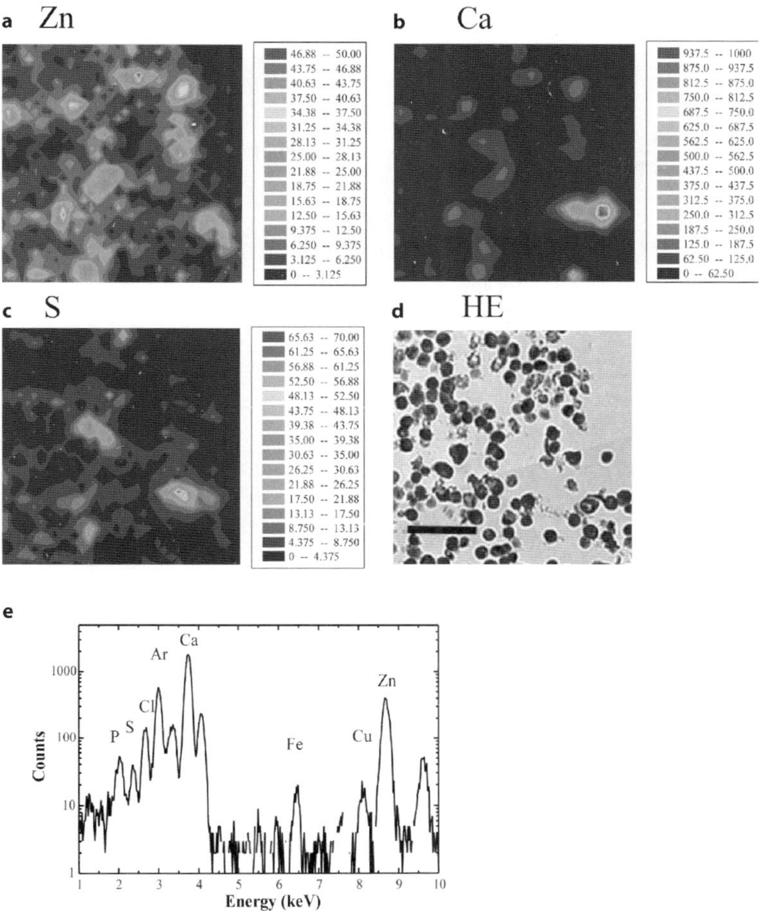

Fig. 8.10a–e. Elemental images of Zn, Ca and S in the cell culture of DV145. The scanning area are $175 \times 160\ \mu m^2$. The XRF images have matrices of 36×33 pixels of $5\ \mu m$ resolution and the measurement time was $3\ s$ per point. The color scales at the right side of the images show X-ray fluorescence intensity. The scale bar is $50\ \mu m$

Quantification results of these cell lines are shown in Table 8.3. The average of densities of zinc in DU145, LNCaP and PC-3 cells are 1,068, 49 and 682 ppm, respectively. It should be noted that the average of the density is calculated in the cells containing zinc, not in the cells which contains zinc under background level. The density of Zn in DU145 and PC-3 are higher than that in LNCaP, indicating that aggressive cancer cells such as DU145 and PC-3 have high density of Zn. The result of the Ca densities in DU145, PC-3 and LNCaP shows the same tendency as those of Zn.

According to the results of the measurements using the homogenization method at bulk tissue level, it is well known that the density of zinc in prostate cancer tissues is at a lower level than that of control tissues or BPH

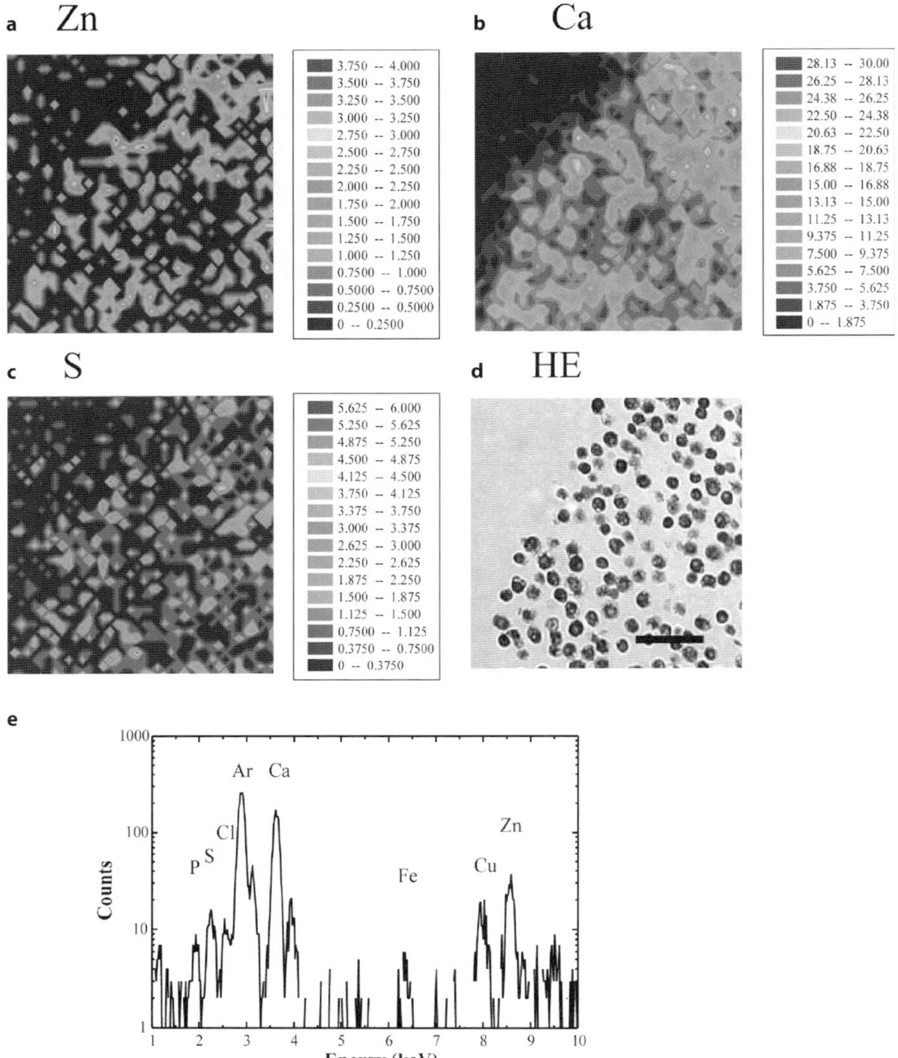

Fig. 8.11a–e. Elemental images of Zn, Ca, S and Fe in the cell culture of LNCaP. The scanning area are $200 \times 200\,\mu m^2$. The XRF images have matrices of 40×40 pixels of $5\,\mu m$ resolution and the measurement time was $3\,s$ per point. The color scales at the right side of the images show X-ray fluorescence intensity. The scale bar is $50\,\mu m$

tissues [1, 2]. Using a synchrotron radiation microbeam with high brilliance, one can determine quantitative information without isolation or purification of the target element, allowing discussion of the density of trace metallic elements at the single cell level. In the first part of the study, Fujisawa et al.

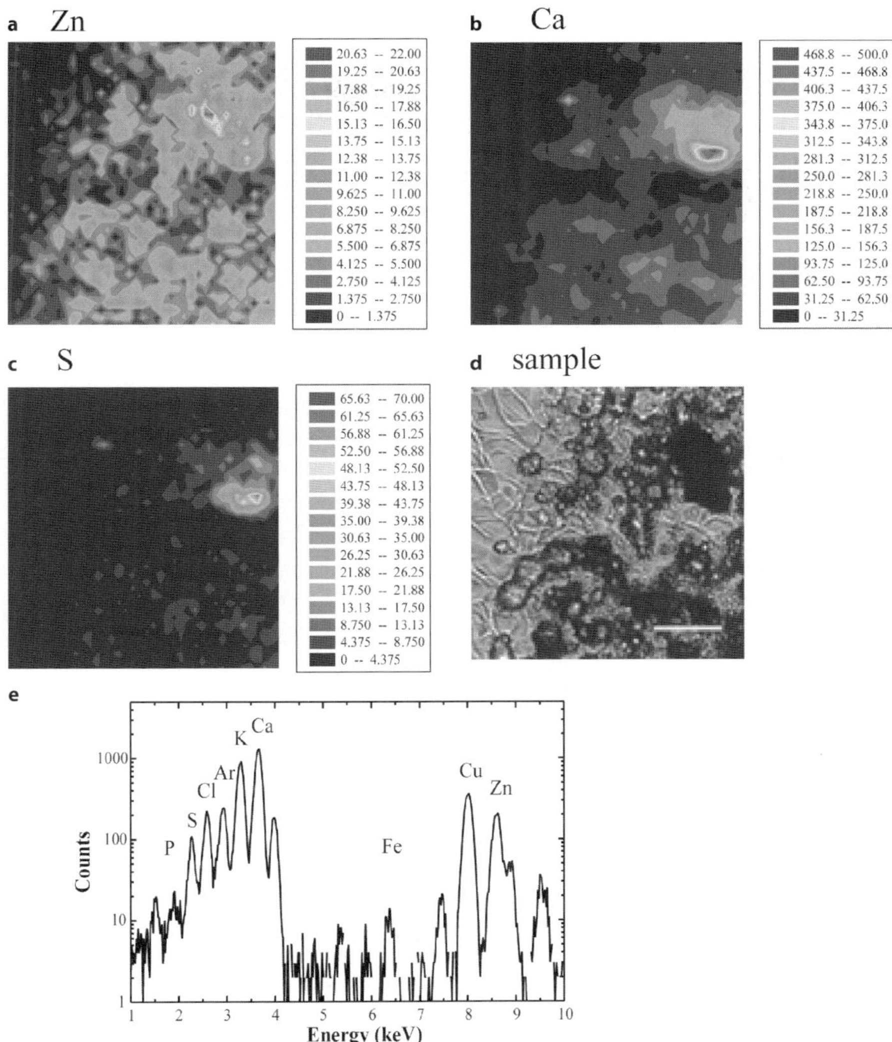

Fig. 8.12a–e. Elemental images of Zn, Ca, S and Cu in the cell culture of PC-3. The scanning area are $200 \times 200\ \mu m^2$. The XRF images have matrices of 40×40 pixels of 5 μm resolution and the measurement time was 3 s per point. The color scales at the right side of the images show X-ray fluorescence intensity. The scale bar is 50 μm

(2002) investigated the differences of the density and the distribution of zinc at the single cells level in cancerous and normal tissues. In comparison with the gland part of the prostate tissue, the density of Zn in the normal tissue (epithelial cells, 158–474 ppm) was higher than that in the cancer tissue (adenocarcinoma, 89–221 ppm). In the cancer tissues, however, high levels of Zn (max 713 ppm) were detected at some points in the stroma of the prostate.

Table 8.3. The density of Zn and Ca contained in the prostate tissues (in ppm). The values are the average of data at n points

	P	S	Cl	Ca	Fe	Cu	Zn	n
DU145	12,288	8,760	11,255	82,487	186	71	1,068	8
LNCaP	2,746	2,380	983	3,839	35	73	89	5
PC3	3,806	8,750	12,913	28,356	77	825	682	4

The differences in the average density of Zn in the stroma were not so significant between cancer cells (44–193 ppm) and normal cells (101–180 ppm). In comparison with normal tissues, the densities of zinc were decreased inside certain cancerous cells and in part of stroma. However it should be noted that at certain points of stroma, accumulation of very high level of Zn can be observed (Fig. 8.1b).

SR enables simultaneous multi-element measurement, in turn allowing assessment of the balance and the equilibrium of elements in the cells as a new approach to cell microbiology. In their study, Fujisawa et al. investigated the correlation between the density of Zn and Ca in the cells in prostate tissues. In the second part of the study, significant differences of distribution of correlation could be seen between adenocarcinoma and stroma in the cancerous tissues. And the average of the zinc density in the stroma was lower than that of adenocarcinoma, indicating that zinc reduction in the tissue with prostate tissue occurs in the stroma rather than in the adenocarcinoma. The decrease of zinc level in the prostate cancer tissue has been known for a long time; however, the differences of the reduction level between stroma and adenocarcinoma has not been referred. The result may suggest the new insight of zinc metabolism in the prostate cancer tissues.

The study using the cultured cells with prostate cancer also shows the interesting relationship between zinc and malignancy. The aggressive cancerous cells such as DU145 and PC-3 contained higher levels of zinc. It is reported that the zinc uptake increases in PC-3 and LNCaP when the cells are stimulated by hormone [13]. It is supposed that the density of zinc is dependent on the environment of the cells. Then it is important to analyze the metabolism alterations of zinc and other elements in these cell lines when the environment (the density of hormone or zinc etc.) is changed. It would be of interest to analyze the zinc metabolism in these cancerous cells under the various environments.

8.1.3 Summary

Metallic elements and their organic compounds have dynamic regulatory functions in cells. The role of zinc and its possible causal effects in prostate cancer has attracted attention in recent years. Fujisawa et al. (2002) applied X-ray fluorescence (XRF) spectroscopy using a synchrotron radiation

(SR) microbeam to the cancerous tissues of the human prostate and the cultured cells in order to investigate the differences of the density and the distribution of zinc in these tissues. In the first part, XRF analysis was applied to the cancerous and normal tissue with human prostate. Zinc was detected in both the cancerous and the normal tissues. In the cancerous tissues, the density of zinc was $89 \sim 221$ ppm in the adenocarcinoma and $44 \sim 713$ ppm in the stroma. In the normal tissues, zinc was detected $158 \sim 474$ ppm in the epithelial cells and $101 \sim 180$ ppm in the stroma. Calcium and other trace metallic elements were also detected in the cancerous and the normal tissues. In the second part, significant differences of distribution of correlation could be seen between adenocarcinoma and stroma in the cancerous tissues. And the average of the zinc density in the stroma is lower than that of adenocarcinoma, indicating that zinc reduction in the tissue with prostate tissue occurs in the stroma rather than in the adenocarcinoma. In the final part, the XRF study using the cultured cells with prostate cancer was performed in order to investigate the relationship between the hormone resistance and the zinc. The aggressive cancerous cells such as DU145 and PC-3 contained higher levels of zinc.

8.2 Application in the Development of New Implant Material

8.2.1 Introduction

The polymers, metals, and ceramics used as implant materials are inserted in the human body every year. The implant materials are in contact with the human tissues directly. If the mechanical strength is required for the implants, it is essential to employ metals for the implant. However, when metals are utilized as the implant materials, their corrosion and the metal ion release from the implant become serious problems. Therefore, there have been extensive efforts to achieve better biocompatibility, corrosion resistance, and wear resistance for the implant parts. At present, commercially pure titanium (Cp-Ti), titanium-alloys (Ti-6Al-4V), Co-Cr alloys and stainless steel are often utilized as the dental and orthopedic implants. It is expected that a thin oxide film of a few nm thickness is immediately formed on the metal surface that acts as a protective barrier. Nevertheless, Ektessabi et al. detected titanium ion-release and measured the distribution of the released elements from titanium and titanium-alloy implants using point analysis technique with microbeam PIXE [14–18]. Some other researchers have also reported on the metal ion release from the implant materials, such as the hip prosthesis and dental biomaterials [19, 20]. It is plausible that the individual elements included in the tissues in the cell level are maintained within a certain range of concentrations. If the densities of some elements increase or decrease excessively, these elements may have toxic effects on the tissues. Therefore, it is supposed that the metal ions or particles released from the implant materials which were inserted in the human body may give the injurious effects to the human tissues. At present, many of the implant materials inserted in the human body are alloys that include some toxic metal elements. It is very important to identify the elements that are released as the corrosion products and their chemical states after the interactions with the cells. Furthermore, it is necessary to investigate the interactions between the cells and metal elements in order to improve the safety and reliability of the implant materials and contribute to development of the new metallic implant materials.

In order to investigate the reactions of the cells against the corrosion products released from the implant, distribution and chemical state analyses must be carried out using microbeams with diameter of micron or sub-micron order. The metallic corrosion products are very small in quantity. The quantification and the chemical state analysis of the dispersed metal elements must be performed at trace element level (a few ppm).

In a study by Kitamura et al. (2000), SR-XRF and XAFS analysis were employed to measure the trace elements in a localized area in human tissues around the total hip replacement. SR-XRF analysis was utilized to investigate the elemental distribution patterns and the quantification of the included elements in the tissues samples. XAFS analysis was applied to investigate the chemical state of the metal elements distributed into the tissues.

8.2.2 Clinical Background and Sample Preparation

In the study, Kitamura et al. used specimens from the case of female and male patients, 55 (case 1) and 60 (case 2) years of age diagnosed with arthrosis. The patients had total hip replacements with a hydroxy-apatite (HAp) coated prosthesis. The implant consisted of a stem and a metal backing made of Ti-6Al-4V, an implant head made of stainless steel, and a polyethylene (PE) cup. Both the stem and the metal backing had a plasma-sprayed HAp surface coating with a thickness of 155 μm. Accelerating PE wear was diagnosed, leading to re-operation 5.4 and 5.9 years after insertion, respectively. These hips functioned painlessly until a few months before re-operation. At revision, one could observe excessive wear, i.e., the femoral head had created a wear hole through the PE-inlay and the steel was fretting directly on the Ti-alloy made backing. Stained ground sections (10 μm thick) were prepared for light microscopic observations as well as thin (50 μm) stained ground sections for SR-XRF spectroscopy. A schematic drawing of the typical hip replacement prosthesis is shown in Fig. 8.13 and photographs of the specimens are shown in Fig. 8.14.

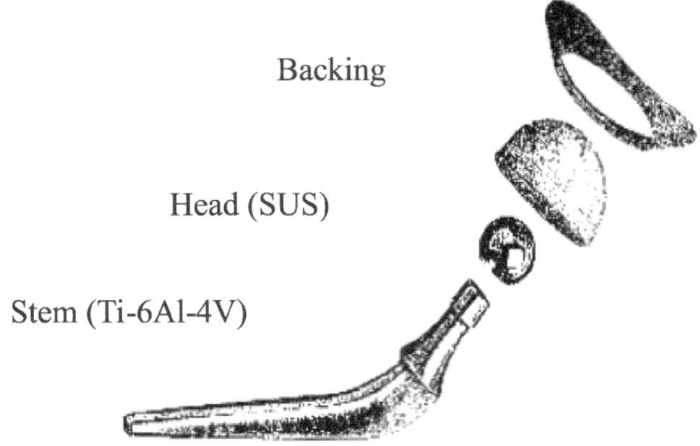

Fig. 8.13. Schematic drawing of a typical hip replacement prosthesis

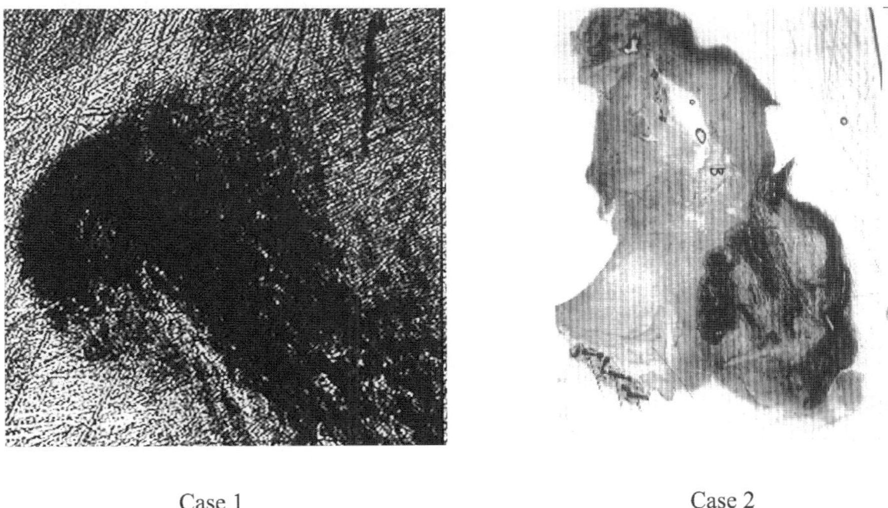

Fig. 8.14. Photographs of the specimens for case 1 and 2

8.2.3 Results

8.2.3.1 Case 1

In the light microscopic photograph of the sample from case 1, shown in the Fig. 8.14, it can be observed that there are black parts distributed into the tissues around the total hip replacement. It is assumed that they are the metal particles generated by the mechanical frictions between the head and backing of the hip replacement prosthesis dispersed into the tissues.

In order to make the elemental distributions of the black parts clear, SR-XRF imaging technique was employed. All of the experimental results measured from the case 1 sample were obtained at BL39XU of SPring-8. SR-

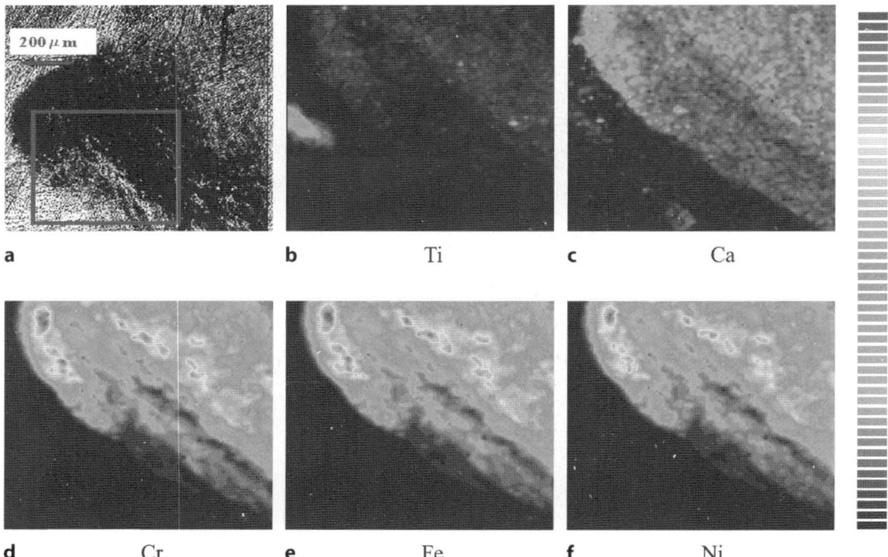

Fig. 8.15. (a) Optical microscopic photograph of the tissues around a failed hip replacement prosthesis (case 1). (b)–(f) Elemental maps of (b) Ti, (c) Ca, (d) Cr, (e) Fe and (f) Ni. All images are 50×50 pixels of $10\,\mu m$ resolution and the measurement time was $1\,s/pixel$. The scale bar on the right shows the count of the X-ray intensity. The range of intensity is 210 for Ti, 84 for Ca, 790 for Cr, 3,300 for Fe, and 490 for Ni

XRF analysis was done within the area surrounded by the red square part shown in Fig. 8.15a. The incident X-ray excitation energy was 10.5 keV and beam had 10 µm in diameter. The elemental distribution patterns obtained by XRF are shown in Fig. 8.15b–f. It can be concluded that the metal particles (stainless steel and Ti alloy) were distributed into the tissues around the total hip replacement, because of the fact that Fe, Cr, Ni and Ti were detected in the area shown in Fig. 8.15a. The fluorescent X-ray signals of Fe, Cr and Ni were much higher than those of Ti in the tissues. This suggests that a number of particles of the stainless steel (SUS316) were generated due to the friction between the stem and the backing of the hip replacement prosthesis.

The result for Ca distribution is shown in Fig. 8.15c, where it can be seen that the pattern is similar to that of Fe, Cr and Ni. The surface of the stem and backing was coated with HAp, but this distribution pattern of Ca did not originate from the particles of HAp due to the fact that the fluorescent X-ray intensities of P and Ca detected in the area shown in Fig. 8.15 are different from the composition of HAp. It is likely that the distribution of Ca may originate from within the cell.

XRF spectra were measured at the po ints shown in Fig. 8.16. The spectrum measured at the point ① is shown in Fig. 8.17. The main elements detected in the energy range from 1 keV to 10 keV were P, S, Ca, Ti, V, Cr,

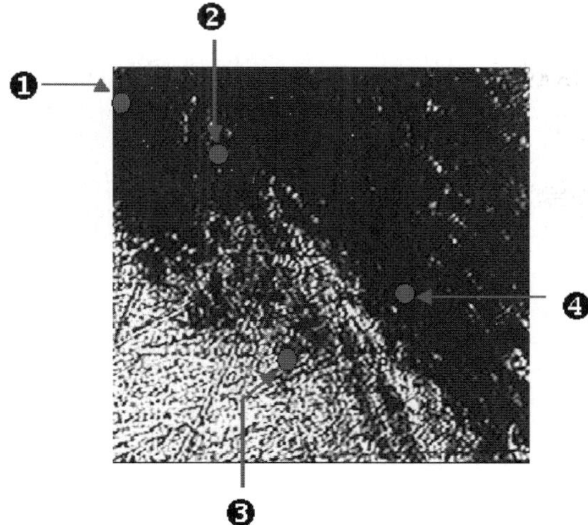

Fig. 8.16. Photograph of the specimen of the case 1 showing the points were XRF spectra were measured

Fe, Ni, Cu and Zn. The peak areas were integrated and the background signals subtracted. The values of these integrated counts are shown in Table 8.4. These values are corrected and the composition ratios of Cr, Fe and Ni are calculated and shown in Table 8.5. The concentration of Fe at point ③ was higher than that of the other points. The fluorescent X-ray signals of Cr, Fe and Ni measured at points ①, ② and ③ are very high, leading to conclusion that the black area shown in Fig. 8.15a contains the particles of stainless steel. But, the compositions of Cr, Fe and Ni in the particles were not the same as those of SUS316L utilized to make the stem of the hip replacement prosthesis. In order to find an explanation for this discrepancy, a stainless steel (SUS304) sample was analyzed for the composition of Cr, Fe and Ni near the surface by using auger electron spectroscopy (AES) analysis (data not shown). The compositions of Cr, Fe and Ni near the surface were different from those bulk composition. Therefore, it is plausible that the stainless steel particles produced by the mechanical friction in the prosthesis sample have different composition for Cr, Fe and Ni from the specification of the material. The result shown in Table 8.5, thus, reflects the changes of the metal composition near the surface of the stainless steel.

The relative concentration of Fe measured at the point ③ was different from the other points. This result will be discussed after the chemical state analysis.

In order to investigate the chemical state of the Fe, Kitamura et al. employed XANES analysis in the same area shown in Fig. 8.15a. Fe, FeO, Fe_2O_3 and $FeCl_2 \cdot nH_2O$ were used as the reference samples. All reference samples were measured by transmission mode. At points ① to ④, XANES spectra were obtained by the fluorescence mode. The results of the XANES spec-

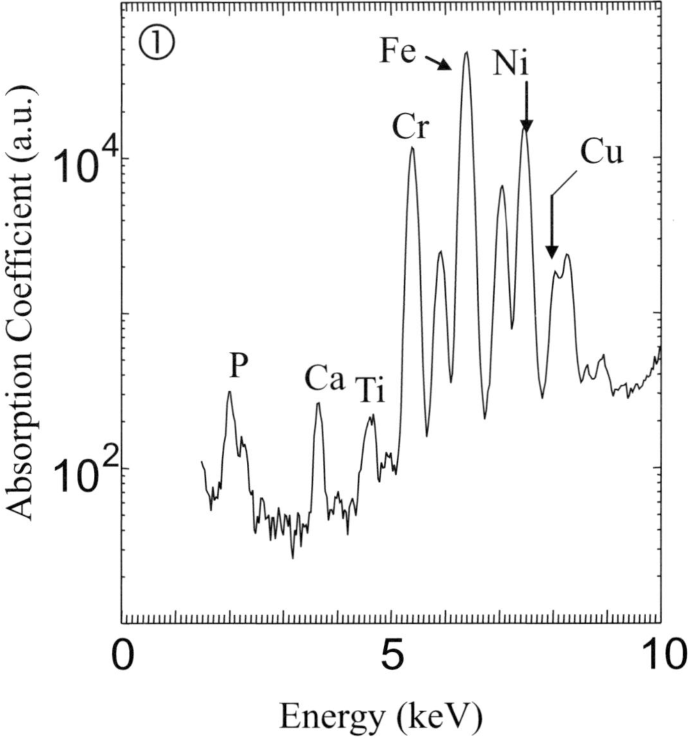

Fig. 8.17. Spectrum measured at the point ① shown in Fig. 8.16

Table 8.4. Values of the integrated counts from the spectrum shown in Fig. 8.17

	Ca	Ti	Cr	Fe	Ni
No.1	1,512.2	1,293.2	71,044	302,240	103,790
No.2	1,058.9	784.4	40,661	177,490	59,483
No.3	633.4	0	2,200.2	10,579	33,557
No.4	804.3	610.76	27,315	122,170	40,541

Table 8.5. Table of the composition ratios of the metallic elements from the spectrum shown in Fig. 8.17

	Ca	Ti	Cr	Fe	Ni
No.1	0.0737	0.0413	1.4	3.99	0.916
No.2	0.0485	0.0227	0.726	2.13	0.476
No.3	0.0276	0	0.0374	0.121	0.0256
No.4	0.0326	0.0156	0.431	1.29	0.287

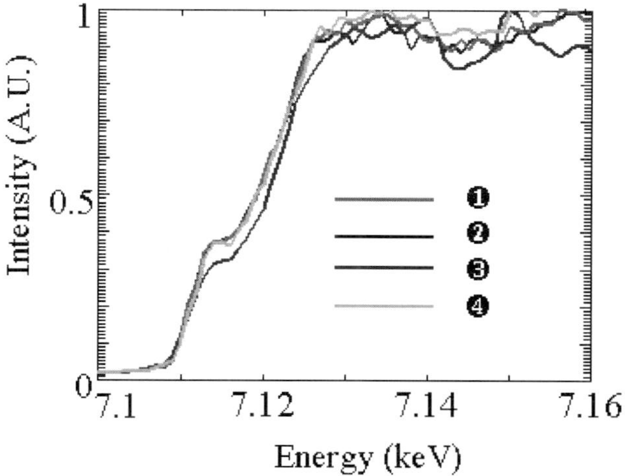

Fig. 8.18. XANES spectra of Fe at the measured points shown in Fig. 8.16

tra are shown in Fig. 8.18. The measurement times of the spectra for points ① and ② were 10 seconds and for points ③ and ④ were 20 seconds.

As can be seen in Fig. 8.18, the XANES spectra of the measured points ①, ② and ④ are almost identical. However, defining the absorption edge position as the half height position of the maximum intensity, the absorption edge of the spectrum measured at the position ③ shifted to high energy compared with those for points ①, ② and ④. For better comparison, the spectra of points ① and ③ were compared with the XANES spectrum of a reference sample (Fe) (data not shown). There were no differences between the absorption edge position of the spectrum of Fe and the measured point ①. The little differences of the shape of the spectra between Fe and the measured point ① may be due to the difference of the crystal structure between Fe and stainless steel. The crystal structure of Fe is *b.c.c.* and that of the stainless steel (SUS316L) is *f.c.c.* state. Therefore, it is plausible that the chemical states of Fe at points ①, ② and ④ are virtually the same as that for stainless steel. It is assumed that the stainless steel particles dispersed from the friction parts of the hip replacement prosthesis are mainly confined in the black tissues as shown in Fig. 8.15a. However, the chemical state of Fe around point ③ is different from that in the black tissues. It is assumed that the corrosion from the stainless steel particles occurred in the area around the measured point ③. Furthermore, it is probable that the selective release of Fe from the stainless steel particles may occur near the areas that include the mechanical friction particles.

In order to investigate the chemical state of chromium, XANES spectrum was obtained at the point ④. The XANES spectra from Cr, Cr_2O_3 standards and point ④ are shown in Fig. 8.19. There is only a little difference between the shapes of the spectra from Cr and measured point ④. The little change

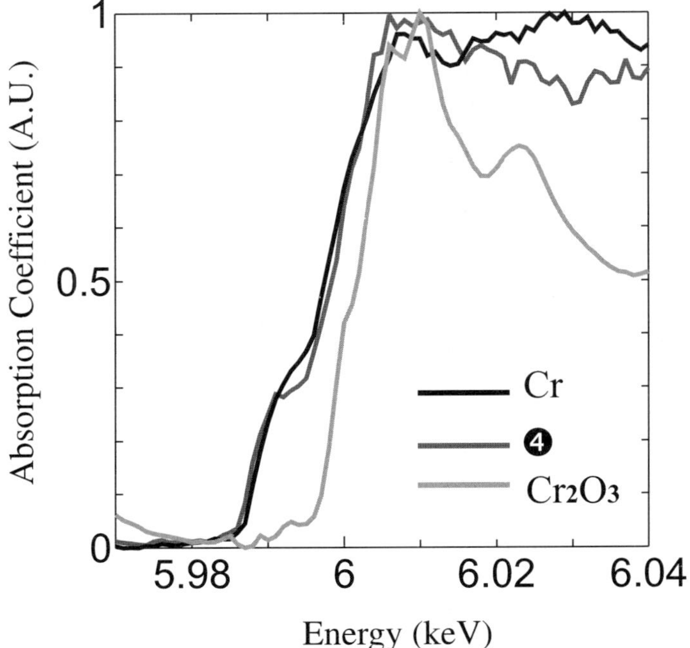

Fig. 8.19. The XANES spectra of Cr, Cr_2O_3 and measured point ④

of the shape of the spectrum at the point ④ may be reflect the differences of the crystal structures between Cr and Cr within the stainless steel particles and Cr_2O_3 formed on the surface of the particles.

8.2.3.2 Case 2

The sample named case 2 is from a male patient of 60 years of age diagnosed with arthrosis. In the tissues around the total hip replacement, there are many black areas similar to those observed in case 1. From the results of SR-XRF imaging analysis, it is clear that the metal particles originating from the friction parts of the hip replacement prosthesis are distributed into the tissues. The detailed elemental images of Cr and Fe within the tissues around the hip replacement prosthesis are shown in Fig. 8.20. This area of the tissue is expected to contain many of the dispersed stainless steel particles. The scanning area was $250\,\mu m \times 250\,\mu m$. These images were 50×50 pixels of $5\,\mu m$ resolution. In this figure, the intensities of Fe is reduced by 4.5. Under this condition, the elemental distribution patterns of Fe and Cr within the tissues are displayed identically.

The images of Cr and Fe within from a different area of the tissue are shown in Fig. 8.21. The scanning area was $100\,\mu m \times 100\,\mu m$. These images were 100×100 pixels of $1\,\mu m$ resolution. The same reduction (4.5) of Fe

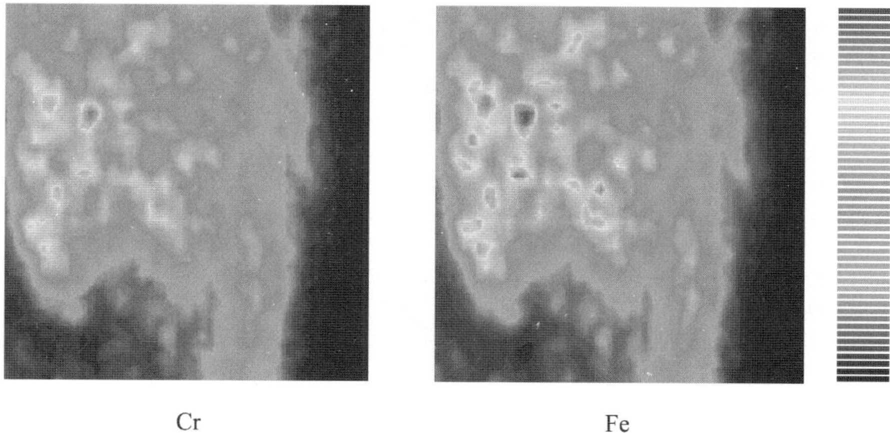

Fig. 8.20. Detailed images of Cr and Fe within the tissues around the hip replacement prosthesis of the case 2

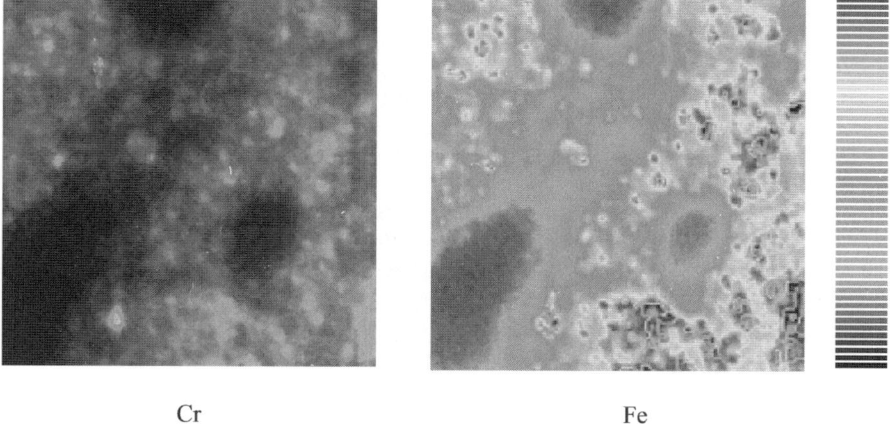

Fig. 8.21. Detailed elemental images of Cr and Fe from a different area of the tissue of the case 2 sample

intensity is used. In this case, the elemental distribution patterns of Fe and Cr within the tissues are different from each other. The distribution pattern of Fe is extended as compared to that of Cr. It is plausible that the extended Fe distribution reflects selective Fe release from the stainless steel particles, due to long period of the exposure within the human tissues.

8.2.4 Discussion

The results of SR-XRF imaging and XANES analyses suggest that selective iron release from the stainless steel particles may occur in the tissues around the total hip replacement due to the long period of exposure after the emission of the mechanical friction particles originating from the friction point between the stem and backing of the hip replacement prosthesis. It is necessary to clarify the chemical state of iron released from the implant materials as the corrosion products. However, the concentrations of the corrosion products from the implant materials are too low for unambiguous interpretation and the structures of iron included in the corrosion material are extremely complex. Other implant materials, such as titanium and Co-Cr alloy, also have the problems of the friction and corrosion. By using the microbeam from the synchrotron radiation source, ultra-trace element analysis is realized to investigate the elemental distributions and the chemical states of the elements within the biological samples. It is necessary to do further investigations into the problems of cytotoxicity of the corrosion materials after the interaction with the cells.

References

1. F. Gyorkey, K.W. Min, J.A. Huff, P. Gyorkey, *Cancer Res.,* **1967**, 27, 1349.
2. F.K. Habib, M.K. Mason, P.H. Smith, S.R. Stitch, *Br. J. Cancer,* **1979**, 39, 700.
3. J.E. Baldwin, H. Krebs, *Nature,* **1981**, 291, 381.
4. L.C. Costello, Y.Y. Liu, R.B. Franklin, M.C. Kennedy, *J. Biol. Chem.,* **1997**, 272, 28875.
5. J.E. Cooper, I. Farid, *J. Surg. Res.,* **1963**, 3, 112.
6. L.C. Costello, R.B. Franklin, *Prostate,* **1994**, 25, 162.
7. L.C. Costello, R.B. Franklin, *Prostate,* **1998**, 35, 285.
8. K. Kimura, M. Markowski, C. Bowen, E.P. Gelmann, *Cancer Res.,* **2001**, 61, 5611.
9. S.J. Eom, E.Y. Kim, J.E. Lee, H.J. Kang, J. Shim, S.U. Kim, B.J. Gwag, E.J. Choi, *Mol. Pharmacol.,* **2001**, 59, 981.
10. J.S. Horoszewicz, S.S. Leong, E. Kawinski, J.P. Karr, H. Rosenthal, T.M. Chu, E.A. Mirand, G.P. Murphy, *Cancer Res.,* **1983**, 43, 1809.
11. P. Angelloznicoud, M. Binoux, *Endocrinology,* **1995**, 136, 5485.
12. D. Peng, Z. Fan, Y. Lu, T. DeBlasio, H. Scher, J. Mendelsohn, *Cancer Res.,* **1996**, 56, 3666.
13. L.C. Costello, Y.Y. Liu, J. Zou, R.B. Franklin, *J. Biol. Chem.,* **1999**, 274, 17499.
14. K. Donath, *"Preparation of Histologic Sections",* **1995**, Exakt-Kulzer Publication, Norderstedt.
15. A.M. Ektessabi, T. Otsuka, Y. Tsuboi, K. Yokoyama, K. Fujii, T. Albrektsson, L. Sennerby, C. Johansson, *Int. J. PIXE,* **1994**, 4, 81.
16. A.M. Ektessabi, A. Wennerberg, *Int. J. PIXE,* **1995**, 5, 145.
17. J.C. Wataha, C.T. Hanks, Z. Sun, *Dent. Mater.,* **1995**, 11, 239.
18. D. Granchi, G. Ciapetti, S. Stea, L. Savarino, F. Filippini, A. Sudanese, G. Zinghi, L. Montanaro, *Biomaterials,* **1999**, 20, 1079.
19. K. Zaman, H. Ryu, D. Hall, K. O'Donovan, K. Lin, M.P. Miller, J.C. Marquis, J.M. Baraban, G.L. Semenza, R.R. Ratan, *J. Neurosci.,* **1999**, 19, 9821.
20. K. Abreo, F. Abreo, M.L. Sella, S. Jain, *J. Neurochem.,* **1999**, 72, 2059.

Index